ILLUSIONISM

AS A THEORY OF CONSCIOUSNESS

Edited by
Keith Frankish

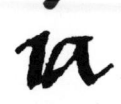

ia

Imprint-academic.com

Copyright © Imprint Academic Ltd., 2017

The moral rights of the authors have been asserted.
No part of this publication may be reproduced in any form
without permission, except for the quotation of brief passages
in criticism and discussion.

Published in the UK by
Imprint Academic, PO Box 200, Exeter EX5 5YX, UK

Distributed in the USA by
Ingram Book Company,
One Ingram Blvd., La Vergne, TN 37086, USA

ISBN 9781845409579

A CIP catalogue record for this book is available from the
British Library and US Library of Congress

Contents

5	About Authors	
9	Editorial Introduction	*Keith Frankish*

Target Paper

11	Illusionism as a Theory of Consciousness	*Keith Frankish*

Commentaries

40	Illusionism's Discontent	*Katalin Balog*
52	Delusions of Consciousness	*Susan Blackmore*
65	Illusionism as the Obvious Default Theory of Consciousness	*Daniel C. Dennett*
73	Illusionism and Givenness	*Jay L. Garfield*
83	Is Realism about Consciousness Compatible with a Scientifically Respectable Worldview?	*Philip Goff*
98	Consciousness Engineered	*Michael S.A. Graziano*
116	Redder than Red: Illusionism or Phenomenal Surrealism?	*Nicholas Humphrey*
124	The Hardest Aspect of the Illusion Problem — and How to Solve it	*François Kammerer*
140	Meta-Illusionism and Qualia Quietism	*Pete Mandik*
149	A Split-Brain Perspective on Illusionism	*Nicole L. Marinsek & Michaeal S. Gazzaniga*
160	The Illusion of Illusionism	*Martine Nida-Rümelin*
172	Illusionism and Anti-Functionalism about Phenomenal Consciousness	*Derk Pereboom*
186	Against Illusionism	*Jesse Prinz*

197	Taking Consciousness Seriously — as an Illusion	*Georges Rey*
215	Illusionism and the Epistemological Problems Facing Phenomenal Realism	*Amber Ross*
224	Phenomenal Consciousness, Defined and Defended as Innocently as I Can Manage	*Eric Schwitzgebel*
236	What is at Stake in Illusionism?	*James Tartaglia*

Reply to Commentators

256	Not Disillusioned: Reply to Commentators	*Keith Frankish*

ABOUT AUTHORS

Katalin Balog moved to the USA from Budapest to study philosophy at Rutgers University, where she got her PhD in 1998. She taught philosophy at Cornell University, and then Yale University from 1998–2010. In 2010 she moved to Rutgers University/Newark where she is an Associate Professor of Philosophy. Her primary areas of research are the philosophy of mind and metaphysics. The problems that interest her most, the nature of consciousness, the self, and free will, lie at their intersection.

Susan Blackmore is Visiting Professor at the University of Plymouth and a freelance writer and lecturer. She blogs for the *Guardian* and *Psychology Today*, is a TED lecturer, and frequent contributor and presenter on radio and television. She has a BA in Physiology and Psychology from Oxford and a PhD in Parapsychology from Surrey University. Her books include *Dying to Live* (on NDEs, 1993), *In Search of the Light* (autobiography, 1996), *The Meme Machine* (1999), *Conversations on Consciousness* (2005), *Zen and the Art of Consciousness* (2011), and *Consciousness: An Introduction* (2011).

Daniel C. Dennett is University Professor and Austin B. Fletcher Professor of Philosophy at Tufts University, and Co-director of the Center for Cognitive Studies. He is the author of *Consciousness Explained*, *Sweet Dreams*, and many scholarly articles on the science of consciousness.

Keith Frankish is Visiting Research Fellow at The Open University UK, and Adjunct Professor with the Brain and Mind Programme in Neurosciences at the University of Crete, Greece. He is the author of *Mind and Supermind* (2004) and *Consciousness* (2005), as well as numerous articles and book chapters. He is co-editor of *In Two Minds: Dual Processes and Beyond* (with Jonathan Evans, 2009), *New Waves in Philosophy of Action* (with Jesús Aguilar and Andrei Buckareff, 2010), *The Cambridge Handbook of Cognitive Science* (with William Ramsey, 2012), and *The Cambridge Handbook of Artificial Intelligence* (with William Ramsey, 2014). His research interests include the nature of phenomenal consciousness, the psychology of belief, and dual-process theories of reasoning.

Jay L. Garfield is Doris Silbert Professor in the Humanities and Professor of Philosophy, Logic, and Buddhist Studies at Smith College, Visiting Professor of Buddhist Philosophy at Harvard Divinity School,

Professor of Philosophy at Melbourne University, and Adjunct Professor of Philosophy at the Central University of Tibetan Studies. His research addresses topics in the foundations of cognitive science and the philosophy of mind; the history of Indian philosophy; topics in ethics, epistemology, and the philosophy of logic; the philosophy of David Hume; methodology in cross-cultural interpretation; and topics in Buddhist philosophy. Recent books include *Minds Without Fear: Philosophy in the Indian Renaissance* (with Nalini Bhushan, 2017), *Dignāga's Investigation of the Percept: A Philosophical Legacy in India and Tibet* (with Douglas Duckworth *et al.*, 2016), *Engaging Buddhism: Why it Matters to Philosophy* (2015).

Michael S. Gazzaniga is the director of the SAGE Center for the Study of the Mind and Professor in the Psychological & Brain Sciences department at UCSB. He is also the president of the Cognitive Neuroscience Institute and the founding director of the MacArthur Foundation's Law and Neuroscience Project and the Summer Institute in Cognitive Neuroscience. Gazzaniga's remarkable research on split-brain patients has expanded our understanding of interhemispheric communication and functional lateralization in the brain.

Philip Goff is Associate Professor of Philosophy at Central European University in Budapest. His research interests are in metaphysics and philosophy of mind, with a special emphasis on the mind–body problem. In his book *Consciousness and Fundamental Reality* (2017, OUP), Goff argues against physicalism, the view that fundamental reality is entirely physical, and in favour of panpsychism, the view that fundamental physical entities are conscious.

Michael S.A. Graziano is Professor of Psychology and Neuroscience at Princeton University. He has worked for thirty years studying sensory processing, multisensory integration, and movement control in the brain. More recently his scientific work has turned to the brain basis of conscious experience. He is an author of numerous books on neuroscience including *Consciousness and the Social Brain*. He has also published literary novels and children's books.

Nicholas Humphrey is a theoretical psychologist who has migrated from neurophysiology, through animal behaviour, to evolutionary psychology and the study of consciousness. He did research on mountain gorillas with Dian Fossey in Rwanda, he was the first to demonstrate the existence of 'blindsight' after brain damage in monkeys, he proposed the celebrated theory of the 'social function of intellect', and he has recently explained the evolutionary basis of the placebo effect. He has held positions at the universities of Oxford and Cambridge, and is now Emeritus Professor at the LSE. Honours include the Martin Luther King Memorial Prize, the Pufendorf medal, and the International Mind and Brain Prize.

François Kammerer is a PhD student in philosophy of mind at the Université Paris-Sorbonne. He is particularly interested in the metaphysics of conscious experience.

Pete Mandik is Professor of philosophy at William Paterson University and author of *This is Philosophy of Mind* (Wiley-Blackwell, 2013).

Nicole L. Marinsek is a PhD student in Dynamical Neuroscience at the University of California, Santa Barbara. She uses behavioural and neuroimaging techniques to research the neural dynamics of inferential reasoning, explanation, and belief updating. She has received several fellowships and awards, including an NSF Graduate Research Fellowship, a SAGE Center Graduate Student Fellowship, and a Doctoral Scholars fellowship. She was also selected to present her research at the Nobel Laureate Meeting in Lindau, Germany.

Martine Nida-Rümelin is full professor for philosophy of mind, language, and the human sciences at the University of Fribourg, Switzerland. Her research area is philosophy of mind with special focus on phenomenal consciousness, identity, and individuality of conscious beings, self-awareness, and agency. She is author of the book *Der Blick von Innen* (Suhrkamp, 2006), a translation of which with additional chapters will soon appear under the title *The View from Inside*.

Derk Pereboom is the Susan Linn Sage Professor of Philosophy and Ethics at Cornell University. His research areas are free will and moral responsibility, philosophy of mind, Kant, and philosophy of religion, and he is the author of *Living without Free Will* (CUP, 2001), *Consciousness and the Prospects of Physicalism* (OUP, 2011), and *Free Will, Agency, and Meaning in Life* (OUP, 2014).

Jesse Prinz is a Distinguished Professor of Philosophy and Director of Interdisciplinary Science Studies at the City University of New York, Graduate Center. His research focuses on the perceptual, emotional, and cultural foundations of human psychology. He is author of *Furnishing the Mind* (MIT, 2002), *Gut Reactions* (OUP, 2004), *The*

Emotional Construction of Morals (OUP, 2007), *Beyond Human Nature* (Penguin, 2012), and *The Conscious Brain* (OUP, 2012).

Georges Rey is Professor of Philosophy at the University of Maryland at College Park. He is the author of numerous articles in the philosophy of mind and cognitive science, and of *Contemporary Philosophy of Mind: A Contentiously Classical Approach* (Blackwell, 1997).

Amber Ross is currently a Post-Doctoral Fellow in the Philosophy Department at the University of Toronto and member of the Network for Sensory Research. Her research is concerned with issues in consciousness and the epistemology of perception, as well as mental states in non-human animals. Her previous position was as a Research Fellow at Tufts University Center for Cognitive Studies under the supervision of Daniel Dennett.

Eric Schwitzgebel is Professor of Philosophy at University of California at Riverside. He is author of *Describing Inner Experience? Proponent Meets Skeptic* (with Russell T. Hurlburt) and *Perplexities of Consciousness*. Recent articles include 'The Crazyist Metaphysics of Mind', 'If Materialism is True, the United States is Probably Conscious', and '1% Skepticism'. He also publishes on moral psychology, Chinese philosophy, and science fiction.

James Tartaglia is Senior Lecturer in Philosophy at Keele University, UK. He is the author of *Rorty and the Mirror of Nature* (2007) and *Philosophy in a Meaningless Life* (2016); editor of *Richard Rorty: Critical Assessments of Leading Philosophers* (2009); and co-editor (with Stephen Leach) of *Mind, Language, and Metaphilosophy: Early Philosophical Papers* (2014) and *Consciousness and the Great Philosophers: What would they have said about our mind–body problem?* (2016). He also leads the band Continuum of Selves; their debut album is entitled Jazz-Philosophy Fusion (2016).

Keith Frankish

Editorial Introduction

The topic of this special issue is the view that phenomenal consciousness (in the philosophers' sense) is an illusion — a view I call *illusionism*. This view is not a new one: the first wave of identity theorists favoured it, and it currently has powerful and eloquent defenders, including Daniel Dennett, Nicholas Humphrey, Derk Pereboom, and Georges Rey. However, it is widely regarded as a marginal position, and there is no sustained interdisciplinary research programme devoted to developing, testing, and applying illusionist ideas. I think the time is ripe for such a programme. For a quarter of a century at least, the dominant physicalist approach to consciousness has been a realist one. Phenomenal properties, it is said, are physical, or physically realized, but their physical nature is not revealed to us by the concepts we apply to them in introspection. This strategy is looking tired, however. Its weaknesses are becoming evident (see, for example, James Tartaglia's contribution to this issue), and some of its leading advocates have now abandoned it. It is doubtful that phenomenal realism can be bought so cheaply, and physicalists may have to accept that it is out of their price range. Perhaps phenomenal concepts don't simply fail to represent their objects as physical but *mis*represent them as phenomenal, and phenomenality is an introspective illusion (this is, in a sense, the physicalist counterpart to the panpsychist approach currently gaining popularity among anti-physicalists — whereas panpsychists think that phenomenal properties are everywhere, illusionists think they are nowhere).

Despite this, it is not easy to persuade people to take illusionism seriously. In part, this is because it is easily caricatured as denying that we have sensations in the everyday sense (it would be more accurate to say that it rejects a certain conception of what sensations are). Moreover, there are some obvious objections to the view. It is often

Correspondence:
Keith Frankish, The Open University, UK.   Email: k.frankish@gmail.com

said that consciousness cannot be an illusion since if it *seems* to us introspectively that we are having a certain conscious experience, then we *are* having it: there is no appearance/reality distinction for consciousness. There is some work to be done, then, just to make illusionism seem worth considering. One aim of this special issue is to do this work.

The issue is focused around a target article, in which I introduce illusionism, sketch the case for the view, and respond to some familiar objections to it. The rest of the issue consists of commentaries on the target article and my reply to them. Many of the commentary authors are sympathetic to illusionism, and their contributions extend and refine the case for the view, exploring it from different perspectives and offering new arguments, insights, and, in some cases, qualifications. The issue is not wholly devoted to defending illusionism, however, and a representative sample of critical perspectives is included as well. Taken together, the issue should give a good sense of the potential of illusionism as a theory of consciousness. I hope it will stimulate interest in the topic and foster a concerted illusionist research programme.

I thank the editor-in-chief of *Journal of Consciousness Studies*, Valerie Hardcastle, for agreeing to this special issue; the journal's managing editor, Graham Horswell, for supervising its production in such a careful and efficient way; all the contributors for their hard work and enthusiasm for the project; and David Chalmers, Daniel Dennett, Nicholas Humphrey, Maria Kasmirli, and Julian Kiverstein for their encouragement and advice. Special thanks are due to Dmitry Volkov of the Moscow Centre for Consciousness Studies for inviting me to take part in an international conference on consciousness and free will held in June 2014 in Greenland. Illusionism was extensively discussed during the conference, and several of the participants have subsequently supplied commentaries for this issue. In addition to thanking them, I'd also like to thank those conference participants whose work is not represented here, mentioning in particular David Chalmers, Patricia Churchland, Paul Churchland, and Andy Clark, and (from the Moscow Centre) Masha Ananina, Artem Besedin, Angelina Dmitrieva, Robert Howell, Anna Kostikova, Anton Kuznetsov, Eugine Loginov, Andrey Mertsalov, and Mikhail Terekhov. Their generosity and inspiration was no illusion.

Heraklion, Crete
October 2016

Keith Frankish

Illusionism as a Theory of Consciousness

Abstract: *This article presents the case for an approach to consciousness that I call* illusionism. *This is the view that phenomenal consciousness, as usually conceived, is illusory. According to illusionists, our sense that it is* like something *to undergo conscious experiences is due to the fact that we systematically misrepresent them (or, on some versions, their objects) as having phenomenal properties. Thus, the task for a theory of consciousness is to explain our illusory representations of phenomenality, not phenomenality itself, and the hard problem is replaced by the* illusion problem. *Although it has had powerful defenders, illusionism remains a minority position, and it is often dismissed as failing to take consciousness seriously. This article seeks to rebut this accusation. It defines the illusionist programme, outlines its attractions, and defends it against some common objections. It concludes that illusionism is a coherent and attractive approach, which deserves serious consideration.*

> So, if he's doing it by divine means, I can only tell him this: 'Mr Geller, you're doing it the hard way.' (James Randi, 1997, p. 174)

Theories of consciousness typically address the hard problem. They accept that phenomenal consciousness is real and aim to explain how it comes to exist. There is, however, another approach, which holds that phenomenal consciousness is an illusion and aims to explain why it *seems* to exist. We might call this *eliminativism* about phenomenal consciousness. The term is not ideal, however, suggesting as it does that belief in phenomenal consciousness is simply a theoretical error,

Correspondence:
Keith Frankish, The Open University, UK.   Email: k.frankish@gmail.com

that rejection of phenomenal realism is part of a wider rejection of folk psychology, and that there is no role at all for talk of phenomenal properties — claims that are not essential to the approach. Another label is 'irrealism', but that too has unwanted connotations; illusions themselves are real and may have considerable power. I propose 'illusionism' as a more accurate and inclusive name, and I shall refer to the problem of explaining why experiences seem to have phenomenal properties as the *illusion problem*.[1]

Although it has powerful defenders — pre-eminently Daniel Dennett — illusionism remains a minority position, and it is often dismissed out of hand as failing to 'take consciousness seriously' (Chalmers, 1996). The aim of this article is to present the case for illusionism. It will not propose a detailed illusionist theory, but will seek to persuade the reader that the illusionist research programme is worth pursuing and that illusionists do take consciousness seriously — in some ways, more seriously than realists do.[2]

1. Introducing illusionism

This section introduces illusionism, conceived as a broad theoretical approach which might be developed in a variety of ways.

1.1. Three approaches to phenomenal consciousness

Suppose we encounter something that seems anomalous, in the sense of being radically inexplicable within our established scientific worldview. Psychokinesis is an example. We would have, broadly speaking, three options. First, we could accept that the phenomenon is real and explore the implications of its existence, proposing major revisions or extensions to our science, perhaps amounting to a paradigm shift. In the case of psychokinesis, we might posit previously unknown psychic forces and embark on a major revision of physics to accommodate them. Second, we could argue that, although the phenomenon is real,

---

[1] When I talk of phenomenal properties not being real or not existing, I mean that they are not *instantiated* in our world. This is compatible with the claim that they exist *qua* properties — a claim which illusionists need not deny.

[2] Defenders of illusionist positions (under various names) include Dennett (1988; 1991; 2005), Hall (2007), Humphrey (2011), Pereboom (2011), Rey (1992; 1995; 2007), and Tartaglia (2013). As Tartaglia notes, Place and Smart also denied the existence of phenomenal properties, which Place described as 'mythological' (Place, 1956, p. 49; Smart, 1959, p. 151).

it is not in fact anomalous and can be explained within current science. Thus, we would accept that people really can move things with their unaided minds but argue that this ability depends on known forces, such as electromagnetism. Third, we could argue that the phenomenon is illusory and set about investigating how the illusion is produced. Thus, we might argue that people who seem to have psychokinetic powers are employing some trick to make it seem as if they are mentally influencing objects.

The first two options are *realist* ones: we accept that there is a real phenomenon of the kind there appears to be and seek to explain it. Theorizing may involve some modest reconceptualization of the phenomenon, but the aim is to provide a theory that broadly vindicates our pre-theoretical conception of it. The third position is an *illusionist* one: we deny that the phenomenon is real and focus on explaining the appearance of it. The options also differ in explanatory strategy. The first is *radical*, involving major theoretical revision and innovation, whereas the second and third are *conservative*, involving only the application of existing theoretical resources.

Turn now to consciousness. Conscious experience has a subjective aspect; we say it is *like something* to see colours, hear sounds, smell odours, and so on. Such talk is widely construed to mean that conscious experiences have introspectable qualitative properties, or 'feels', which determine what it is like to undergo them. Various terms are used for these putative properties. I shall use 'phenomenal properties', and, for variation, 'phenomenal feels' and 'phenomenal character', and I shall say that experiences with such properties are *phenomenally conscious*. (I shall use the term 'experience' itself in a functional sense, for the mental states that are the direct output of sensory systems. In this sense it is not definitional that experiences are phenomenally conscious.) Now, phenomenal properties seem anomalous. They are sometimes characterized as simple, ineffable, intrinsic, private, and immediately apprehended, and many theorists argue that they are distinct from all physical properties, inaccessible to third-person science, and inexplicable in physical terms. (I use 'physical' in a broad sense for properties that are either identical with or realized in microphysical properties.) Again, there are three broad options.

First, there is radical realism, which treats phenomenal consciousness as real and inexplicable without radical theoretical innovation. In this camp I group dualists, neutral monists, mysterians, and those who appeal to new physics. Radical realists typically stress the anomalousness of phenomenal properties, their resistance to functional analysis, and the contingency of their connection to their neural correlates. Second, there is conservative realism, which accepts the reality of phenomenal consciousness but seeks to explain it in physical terms, using the resources of contemporary cognitive science or modest extensions of it. Most physicalist theories fall within this camp, including the various forms of representational theory. Both radical and conservative realists accept that there is something real and genuinely qualitative picked out by talk of the phenomenal properties of experience, and they adopt this as their explanandum. That is, both address the hard problem.[3]

The third option is illusionism. This shares radical realism's emphasis on the anomalousness of phenomenal consciousness and conservative realism's rejection of radical theoretical innovation. It reconciles these commitments by treating phenomenal properties as illusory. Illusionists deny that experiences have phenomenal properties and focus on explaining why they seem to have them. They typically allow that we are introspectively aware of our sensory states but argue that this awareness is partial and distorted, leading us to misrepresent the states as having phenomenal properties. Of course, it is essential to this approach that the posited introspective representations are not themselves phenomenally conscious ones. It would be self-defeating to explain illusory phenomenal properties of experience in terms of real phenomenal properties of introspective states. Illusionists may hold that introspection issues directly in dispositions to make phenomenal judgments — judgments about the phenomenal character of particular experiences and about phenomenal consciousness in general. Or they may hold that introspection generates intermediate representations of sensory states, perhaps of a quasi-perceptual kind, which ground our phenomenal judgments. Whatever the details, they must explain the content of the relevant states in broadly functional terms, and the challenge is to provide an account that explains how real and vivid phenomenal consciousness seems. This is the illusion problem.

---

[3] Although all anti-physicalist theories are radical and all conservative theories physicalist, the radical/conservative distinction does not coincide with the anti-physicalist/physicalist one, since there may be radical physicalist theories.

1.2. Illusionism strong and weak

Illusionism makes a very strong claim: it claims that phenomenal consciousness is illusory; experiences do not really have qualitative, 'what-it's-like' properties, whether physical or non-physical. This should be distinguished from a weaker view according to which some of the supposed *features* of phenomenal consciousness are illusory. Many conservative realists argue that phenomenal properties, though real, do not possess the problematic features sometimes ascribed to them, such as being ineffable, intrinsic, private, and infallibly known. Phenomenal feels, they argue, are physical properties which introspection misrepresents as ineffable, intrinsic, and so on. We might call this *weak illusionism*, in contrast to the strong form advocated here. (It might equally be called *weak realism*.)[4]

On the face of it, weak and strong illusionism are similar. Both hold that experiences have distinctive physical properties that are misrepresented by introspection. There is a crucial difference, however. Weak illusionism holds that these properties are, in some sense, genuinely *qualitative*: there really are phenomenal properties, though it is an illusion to think they are ineffable, intrinsic, and so on. Strong illusionism, by contrast, denies that the properties to which introspection is sensitive are qualitative: it is an illusion to think there are phenomenal properties at all.

We can highlight the difference by introducing the notion of a *quasi-phenomenal property*. A quasi-phenomenal property is a non-phenomenal, physical property (perhaps a complex, gerrymandered one) that introspection typically misrepresents as phenomenal. For example, quasi-phenomenal redness is the physical property that typically triggers introspective representations of phenomenal redness.[5] There is nothing phenomenal about such properties — nothing 'feely' or qualitative — and they present no special explanatory problem. Strong illusionists hold that the introspectable properties of experience are merely quasi-phenomenal ones. But weak

---

[4] For an example of a weak illusionist position, see Carruthers (2000, pp. 93–4, 182–91). For more examples, and discussion, see Frankish (2012), where phenomenal properties in this weakly illusionist sense are dubbed *diet qualia* — in contrast to *classic qualia*, or *qualia max*, which are genuinely ineffable, intrinsic, private, and so on. Compare also Levine's distinction between *modest* and *bold* qualophilia (Levine, 2001, chapter 5).

[5] Extending the soft-drink metaphor, I have dubbed quasi-phenomenal properties *zero qualia* (Frankish, 2012).

illusionists cannot agree. If experiences have only quasi-phenomenal properties, then it would be misleading to say that phenomenal properties are real, just as it would be misleading to say that psychokinetic powers are real if all people can do is create the illusion of having them.

The moral is that if weak illusionism is not to collapse into strong illusionism, then it must employ a concept of phenomenality stronger than that of quasi-phenomenality. Indeed, one motive for advancing the strong illusionist position is to force conservative realists to face up to the challenge of articulating a concept of the phenomenal that is both stronger than that of quasi-phenomenality and weak enough to yield to conservative treatment. I doubt this is possible (see Frankish, 2012) and, if it is not, then radical realism and strong illusionism will be the only options. In what follows, 'illusionism' will always mean *strong illusionism*.

1.3. Some analogies

Illusionists offer various analogies to illustrate their view. Dennett compares consciousness to the user illusions created by the graphical interfaces through which we control our computers (Dennett, 1991, pp. 216–20, 309–14). The icons, pointers, files, and locations displayed on a computer screen correspond in only an abstract, metaphorical way to structures within the machine, but by manipulating them in intuitive ways we can control the machine effectively, without any deeper understanding of its workings. The items that populate our introspective world have a similar status, Dennett suggests. They are metaphorical representations of real neural events, which facilitate certain kinds of mental self-manipulation but yield no deep insight into the processes involved. (Dennett stresses the limits of the interface analogy. There is no internal display for the benefit of a conscious user; the illusion is a product of the limited access relations between multiple non-conscious subsystems and it manifests itself in our personal-level intuitions and judgments about our inner lives.)

Rey cites cases where stabilities in our reactions to the world induce us to project corresponding properties onto the world (Rey, 1995, pp. 137–9). For example, our stable personal concerns and reactions to others lead us to posit stable, persisting selves as their objects. Similarly, Rey suggests, our representations of our own and others' experiences lead us to posit simple mental phenomena corresponding to them. Take pain, for example. We have a 'weak', functional concept of pain, which includes links both to sensory representations of pain encoding information about intensity, apparent location, and so on, and to third-person representations of pain behaviour in others. Reflecting on our own and others' pains, we then develop a 'strong', qualitative concept of pain as the thing that is the immediate object of our pain experiences and the cause of pain behaviour in others.

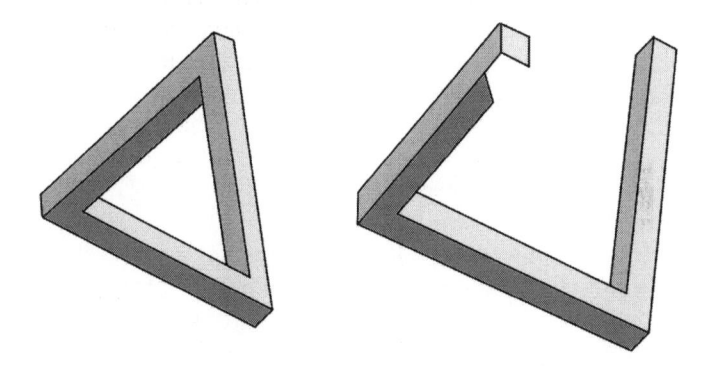

Figure 1. The Penrose triangle and the Gregundrum.

Humphrey compares sensations to impossible objects, such as the Penrose triangle, depicted on the left of Figure 1. Such an object cannot exist in three-dimensional space, but the illusion of it can be created by the object on the right, which Humphrey calls the *Gregundrum*, after its creator Richard Gregory. From most perspectives the Gregundrum appears an ungainly construction, but from just the right angle it looks like a solid Penrose triangle. Consciousness, Humphrey proposes, involves an analogous illusion. Our brains create an 'ipsundrum' — a neural state that appears relatively unremarkable from other perspectives but generates the illusion of phenomenality when viewed introspectively (Humphrey, 2011, chapter 2). Phenomenal consciousness is a '*fiction of the impossible*' (*ibid.*, p. 204) — a magic trick played by the brain on itself. (Talk of illusion should not be taken to indicate a defect in introspection; Humphrey argues that the illusion is highly adaptive; Humphrey, 2006; 2011.)

Pereboom draws a comparison with secondary qualities, such as colours (Pereboom, 2011, pp. 15–40). It is arguable that sensory perception represents colours as properties of external objects resembling the sensations they produce in us. Since objects lack such properties, sensory perception universally misrepresents objects in this respect.

Similarly, Pereboom suggests, introspection may universally misrepresent phenomenal properties as having qualitative natures they do not in fact have. (By 'phenomenal properties' he means the distinctive introspectable properties of conscious experiences, whatever they may be. Phenomenal properties in this sense may be merely quasi-phenomenal.) Pereboom calls this the *qualitative inaccuracy hypothesis*, and he argues that it is an open possibility. If it seems less credible than the parallel hypothesis about secondary qualities, Pereboom suggests, this is because we cannot check the accuracy of introspection, as we can that of perception, by adopting different vantage points, using measuring instruments, and so forth (*ibid.*, p. 23).

These analogies all illustrate the basic illusionist claim that introspection delivers a partial, distorted of view of our experiences, misrepresenting complex physical features as simple phenomenal ones. Sensory states have complex chemical and biological properties, representational content, and cognitive, motivational, and emotional effects. We can introspectively recognize these states when they occur in us, but introspection doesn't represent all their detail. Rather, it bundles it all together, representing it as a simple, intrinsic phenomenal feel. Applying the magic metaphor, we might say that introspection sees the complex sleight-of-hand performed by our sensory systems as a simple magical *effect*. And, as with a conjuring trick, the illusion depends on what the audience does not see as much as what they do. In another analogy, Rey compares our introspective lives to the experience of a child in a dark cinema who takes the cartoon creatures on screen to be real (Rey, 1992, p. 308). The illusion depends on what the child doesn't see — on the fact that their visual system does not register individual frames as distinct images. Cinema is an artefact of the limitations of vision, and, illusionists may say, phenomenal consciousness is an artefact of the limitations of introspection.

The analogy with visual illusions also holds with respect to cognitive penetrability. Forming the theoretical belief that phenomenal properties are illusory does not change one's introspective representations, and one remains strongly disposed to make all usual phenomenal judgments (and perhaps does still make them at some level). As with perceptual illusions, this may indicate that the phenomenal illusion is an adaptive one, which has been hardwired into our psychology. (However, it may be possible to dispel the illusion partially through indirect means, such as meditation and hypnotic suggestion; see, for example, Blackmore, 2011.)

The analogies also indicate some dimensions along which illusionist theories may differ. One concerns the sensory states that are the basis for the illusion. On most accounts, I assume, these will be representational states, probably modality-specific analogue representations encoding features of the stimulus, such as position in an abstract quality space, egocentric location, and intensity. Accounts will differ, however, on the details of their content, functional role, relation to attentional processes, and so on. Theories will also differ as to which properties of these states are responsible for the illusion of phenomenality (their quasi-phenomenal properties). Is introspection sensitive only to the content of sensory states, or are we also aware of properties of their neural vehicles? Do the reactions and associations evoked by our sensory states also contribute to the illusion of phenomenality?

Relatedly, there are questions about our introspective access to our sensory states. Do we have internal monitoring mechanisms that generate representations of sensory states, and if so what sort of representations do they produce? (Are they thoughts about sensory states or perceptions of their neural vehicles?) Are the introspective representations conscious or unconscious? (They are not phenomenally conscious, of course, but they could be conscious in the psychological sense of being globally available.) Are sensory states continually monitored or merely available to monitoring? Is the introspectability of sensory states a matter of internal access and influence rather than internal monitoring?[6] There are many options here and parallels with higher-order representational theories of consciousness, some of which might be reformulated as illusionist ones.

1.4. Outward-looking illusionism?

I characterized illusionism as the view that phenomenal consciousness is an *introspective* illusion, reflecting the widely held view that phenomenal properties are properties of experience. This may be too restrictive, however. Some theorists hold that experience is *transparent*: when we attend to our experiences, we are aware only of properties of their objects. Thus, redness is experienced as a property

---

[6] On the varieties of introspection, see Prinz (2004).

of surfaces, pain as a property of parts of our bodies, and so on (e.g. Harman, 1990; Tye, 1995; 2000). This points to the possibility of an outward-looking illusionism, on which experience misrepresents distal stimuli as having phenomenal properties. Vision, for example, would represent objects as having illusory phenomenal colours as well as real physical colours (for a view of this kind, see Hall, 2007).

This view can be regarded as a variant of standard, inward-looking illusionism, differing principally on where the illusory phenomenal properties are represented as being located. And, like the inward-looking version, it may posit processes of internal monitoring. The illusion of phenomenality may involve a combination of introspection and projection, in which we both misrepresent features of experience as phenomenal and then re-represent these illusory properties as properties of the external world, mistaking complex physical properties of our sensory states for simple phenomenal properties of external objects (Humphrey, 2011, chapter 7). In what follows, I shall focus on the inward-looking form of illusionism, though most points will apply to both.

1.5. Illusionism and grand illusion

Illusionism should be distinguished from the thesis that the visual world is a grand illusion (Noë, 2002). The latter holds that conscious visual experience is far less stable and detailed than we suppose, as is revealed by experiment and careful introspection. Illusionism, by contrast, is a thesis about conscious experience generally and concerns its nature, not its extent. One could hold that the visual world *is* stable and detailed while still claiming that it involves an illusion in the sense discussed here.

Nevertheless, evidence for the grand illusion view, such as the existence of change blindness, does lend support to illusionism. If we regularly overestimate the extent and stability of our conscious visual experience, then it is possible to be under a kind of illusion about one's own phenomenal consciousness. Moreover, as Dennett shows, phenomena such as change blindness undermine familiar intuitions about phenomenal properties, suggesting that our conception of them is incoherent and the properties themselves consequently illusory (e.g. Dennett, 2005, pp. 82–91).

1.6. Illusionism and eliminativism

Does illusionism entail eliminativism about consciousness? Is the illusionist claiming that we are mistaken in thinking we have conscious experiences? It depends on what we mean by 'conscious experiences'. If we mean experiences with phenomenal properties, then illusionists do indeed deny that such things exist. But if we mean experiences of the kind that philosophers *characterize* as having phenomenal properties, then illusionists do not deny their existence. They simply offer a different account of their nature, characterizing them as having merely quasi-phenomenal properties. Similarly, illusionists deny the existence of phenomenal consciousness properly so-called, but do not deny the existence of a form of consciousness (perhaps distinct from other kinds, such as access consciousness) which consists in the possession of states with quasi-phenomenal properties and is commonly mischaracterized as phenomenal. Henceforth, I shall use 'consciousness' and 'conscious experience' without qualification in an inclusive sense to refer to states that might turn out to be either genuinely phenomenal or only quasi-phenomenal. In this sense realists and illusionists agree that consciousness exists.

Do illusionists then recommend eliminating talk of phenomenal properties and phenomenal consciousness? Not necessarily. We might reconceptualize phenomenal properties as quasi-phenomenal ones. Recall Pereboom's analogy with secondary qualities. The discovery that colours are mind-dependent did not lead scientists to deny that objects are coloured. Rather, they reconceptualized colours as the properties that cause our colour sensations. Similarly, we might respond to the discovery that experiences lack phenomenal properties by reconceptualizing phenomenal properties as the properties that cause our representations of phenomenal feels — that is, quasi-phenomenal properties.[7] This could invite confusion, however, given how tightly the notion of phenomenality is bound up with dualist intuitions, and in scientific work it might be wiser to abandon talk of phenomenal properties and phenomenal consciousness altogether.

---

[7] Pereboom suggests that this might involve the unpacking of a conditional structure in phenomenal concepts (Pereboom, 2011, pp. 34–5). As he notes, a given phenomenal property might be reconceptualized either as the neural property that normally causes a representation of the relevant feel or as the higher-order property of being a neural state that could cause a representation of it (*ibid.*).

In everyday life, however, we would surely continue to talk of the feel or quality of experience in the traditional, substantive sense. As subjects of experience, our interest is in how things seem to us introspectively — the illusion itself, not the mechanisms that cause it. Such talk may fail to pick out real properties, but it is not empty or pointless. Consider another analogy. Having watched a performance of *King Lear*, Lucy remarks, 'Lear's anguish in the final scenes was heart-breaking'. What is she talking about? There was (we may suppose) no anguish on stage at all, only the artful illusion of it. And it would be implausible to construe Lucy as referring to the *cause* of this illusion — the actor's words and gestures (quasi-anguish, as it were). The words and gestures were not themselves heart-breaking. The answer, of course, is that Lucy is referring to a fictional agony, entering into the world of the play and responding to the emotions of the characters as if they were real. (And in doing so, we might add, she is not making an error but appreciating the very point of the performance.) Everyday talk about the quality of experience should, I suggest, be construed similarly. Of course, most people do not regard their phenomenology as illusory; they are like naïve theatregoers who take the action on stage for real. But if illusionists are right, then cognitive scientists should treat phenomenological reports as fictions — albeit ones that provide clues as to what is actually occurring in the brain.[8]

1.7. Zombies and what it is like

Are illusionists claiming that we are (phenomenal) zombies? If the only thing zombies lack is phenomenal consciousness properly so called, then illusionists must say that, in this technical sense, we are zombies. However, zombies are presented as creatures very different from ourselves — ones with no inner life, whose experience is completely blindsighted. As Chalmers puts it, 'There is nothing it is like to be a zombie... all is dark inside' (Chalmers, 1996, pp. 95–6). And illusionists will not agree that this is a good description of us. Rather,

---

[8] Compare Dennett's story of the forest god Feenoman (Dennett, 1991, chapter 4). Local tribespeople believe Feenoman is real, but visiting anthropologists treat him as an intentional object, defined by the locals' beliefs, and remain neutral on the question of what lies behind the myth. Dennett recommends that we treat first-person phenomenological reports in the same way, as data for third-person theorizing ('heterophenomenology').

they will deny the equivalence between having an inner life and having phenomenal consciousness. Having the kind of inner life we have, they will say, consists in having a form of introspective self-awareness that creates the illusion of a rich phenomenology.

But aren't phenomenal properties precisely what makes experience *like something*? That is certainly a common way of construing what-it's-like talk, but there is another way. Illusionists can say that one's experiences are like something if one is aware of them in a functional sense, courtesy of introspective representational mechanisms. Indeed, this is a plausible reading of the phrase; experiences are like something for a creature, just as external objects are like something for it, if it mentally represents them to itself. Illusionists agree that experiences are like something in this sense, though they add that the representations are non-veridical, misrepresenting experiences as having phenomenal properties (what-it's-like-ness in the first sense). And in this second sense there *is* something it is like to be a zombie, since zombies have introspective mechanisms functionally identical to our own. When we imagine zombies as being different from us, we are — illegitimately — imagining creatures with different introspective capacities.

It may be objected that we can imagine a creature representing itself as having phenomenal properties while still lacking an inner life. Zombies believe they are phenomenally conscious (in some sense at least; arguably, they lack full-blown phenomenal concepts; Chalmers, 1996; 2003). But — it may be said — this does not give them an inner life like ours. I am not sure this is obvious. Consider the grand illusion view again. Our sense that our visual field is uniformly rich and detailed may be a sort of cognitive illusion, reflecting expectations and assumptions about the information that vision provides, and our sense of having a rich phenomenology might be a similar cognitive illusion. But in any case the illusionist need not claim that the illusion depends solely on the possession of certain propositional attitudes. Rather, they may say, it depends on a complex array of introspectable sensory states, which trigger a host of cognitive, motivational, and affective reactions. If we knew everything about these states, their effects, and our introspective access to them, then, illusionists say, we could not clearly imagine a creature possessing them without having an inner life like ours.

Of course, it is easy to say that. Illusionists need to explain how it can be true. That is, they need to solve the illusion problem. But it would be begging the question against illusionism to assume that it cannot be done.

2. Motivating illusionism

This section motivates illusionism, sketching its advantages over radical realism and conservative realism and then adding some positive arguments in its favour. It does not aim to present a watertight case for illusionism but simply to show that the view has strong attractions.

2.1. Against radical realism

I take it there is a presumption in favour of conservatism in science: we should not make radical theoretical moves if modest ones will do. Of course, when it comes to consciousness many are confident that modest moves *won't* do, but that is what conservative theorists deny. The principle of conservatism should apply with special force, I suggest, when the pressure for radical innovation comes from a parochial, anthropocentric source, such as introspection. Introspection delivers a view of ourselves that is peculiarly vivid and compelling and that seems radically at odds with that of the physical sciences. It *might* give us access to an aspect of reality inaccessible to third-person science. (Though even if it did, it is hard to see how we could develop a science of that aspect.[9]) But it might merely give us an unusual perspective on the same reality — a perspective that is partial and distorted and deceives us into thinking that our experiences are resistant to conservative explanation.

In addition, a conservative approach is much better placed to account for the *psychological significance* of consciousness. By the psychological significance of a mental event, I mean its cumulative cognitive, motivational, emotional, and other psychological effects across various contexts. The common-sense view is that the way our experiences feel has huge psychological significance. Sensations entice us, guide us, move us, warn us, and the memory and anticipation of them are powerful motivators. Not only this, they hugely *enrich* life. As Humphrey stresses, we relish sensation for its own sake, and this relish shapes our behaviour in profound ways

---

[9] For the case against first-person science, see Dennett (1991, chapter 4; 2003; 2005, chapter 6; 2007).

(Humphrey, 2011). But this assumes that experiences affect us in virtue of how they feel. And it is hard for radical theorists to vindicate this assumption. Non-physical properties can have no effects in a world that is closed under causation, as ours appears to be, and the mind sciences show no independent need to refer to exotic physical processes, such as quantum-mechanical ones. The threat of epiphenomenalism hangs over radical theories. Some radical theorists respond by arguing that phenomenal properties are intrinsic to basic physical entities and thus intimately involved in physical causal processes (e.g. Strawson, 2006). However, even if this proposal does dispel the threat (which is doubtful; Howell, 2015), it involves huge profligacy with phenomenal properties and preserves the potency of consciousness only at the cost of making all physical causation phenomenal.

2.2. Against conservative realism

Conservative realism promises to capture the common-sense view of consciousness, accepting the reality of phenomenal properties but identifying them with causally potent, physical properties. However, it is an unstable position, continually on the verge of collapsing into illusionism.

The central problem, of course, is that phenomenal properties seem too weird to yield to physical explanation. They resist functional analysis and float free of whatever physical mechanisms are posited to explain them. (In practice this becomes almost definitional of phenomenal consciousness; any physicalist theory can be rejected as missing out the essential qualitative element.) The arguments are well-known, and I shall not repeat them here.

Many physicalists respond by arguing that our anti-physicalist intuitions arise from the way we conceptualize phenomenal properties in introspection — a tactic known as the *phenomenal concept strategy* (e.g. Hill, 1997; Hill and McLaughlin, 1999; Loar, 1990; Papineau, 2002). The idea is that phenomenal concepts have an especially intimate link to their referents and lack *a priori* connections to physical concepts. (They are typically characterized as either demonstrative, recognitional, or quotational.) This intimacy and isolation, it is argued, give rise to anti-physicalist intuitions, even though phenomenal properties are physical ones. It is doubtful, however, that this really relieves the pressure on conservative realism. For the concepts must still be *phenomenal* ones (Tartaglia, 2013, p. 828). If they are recognitional concepts, for example, they must be recognitional concepts for the *feel* of experiences. The concept of a mere introspectable *something*, which might or might not be qualitative, is not a genuine phenomenal concept, and if we conceptualized the properties of experience in that way, we would not feel any resistance to thinking of them as physical (a bare something might as easily be physical as nonphysical). But if phenomenal concepts refer to feels, then the challenge to conservative realists remains. They must either explain how these feels can be physical or accept that phenomenal concepts misrepresent experience, as illusionists claim.

Looking at proposed reductive explanations themselves, the pressures towards illusionism become even clearer. As noted earlier, most physicalists adopt a weakly illusionist view, denying that phenomenal properties are private, intrinsic, and ineffable and employing the phenomenal concept strategy to explain why they seem so. However, they insist that phenomenal properties are nonetheless real and genuinely qualitative. I have already suggested that this position is problematic. If it is not to collapse into illusionism, then it must employ a notion of phenomenality that is stronger than that of quasi-phenomenality. Phenomenal properties must not merely cause representations of phenomenality but have some genuinely 'feely' aspect to them. And it is unclear what this could be. What phenomenal residue is left, once features such as privacy, intrinsicality, and ineffability have been stripped away (Frankish, 2012)?

In practice, reductive explanations of phenomenality tend to take a covertly illusionist form. They typically identify phenomenal character with some functional property of experience such as possession of a certain kind of representational content or availability to higher-order representation. But in so far as these identifications are plausible, it is, I suggest, because subjects whose experiences had this functional property would be disposed to judge that their experiences had a qualitative dimension, rather than because their experiences really would have such a dimension. In the case of higher-order perception theory, for example, it may be true that perceptual awareness of the physical vehicles of experience would create the sense that experiences have an intrinsic quality. But this is an explanation of quasi-phenomenal properties, not phenomenal properties. There is a conflation of phenomenality with the representation of phenomenality, and thus of realism with illusionism.

Of course, these objections assume that we are seeking an *explanation* of consciousness. Physicalists can resist illusionist pressures if they are content to accept the existence of an *explanatory gap* between phenomenal properties and their neural substrates (e.g. Levine, 2001). Others, however, may prefer an explicable illusion to an inexplicable reality.

It may be objected that illusionism discards one of the major advantages of conservatism, namely that it gives phenomenal properties a causal role. If phenomenal properties are illusory, then they have no causal role after all. Illusionists can reply that they do not deny that phenomenal concepts track causally effective properties; they merely deny the common-sense view of the nature of these properties — that they are qualitative. Or, perhaps more persuasively, they can say that phenomenal properties *are* causally potent, considered as intentional objects. They move us in the same way that ideas, stories, theories, and memes do, by figuring as the objects of our intentional states. In talking of the power of sensation we are talking of the power of certain representational contents.

2.3. For illusionism

The case for illusionism can also be made in a positive way, appealing to explanatory considerations. If phenomenal consciousness is conceived as non-physical, then, as Chalmers notes, there is a simple argument for its being illusory (Chalmers, 1996, pp. 186–7: Chalmers does not endorse the argument, of course, though he acknowledges its force). If people's claims and beliefs about something (God, say, or UFOs) can be fully explained as arising from causes having no connection with the thing itself, then this is a reason for discounting them and regarding the thing as illusory. But it is widely accepted, even by anti-physicalists, that we do not need to appeal to non-physical properties in order to explain our behaviour and the mental processes that cause it, including our assertions and beliefs about our own conscious experiences. Phenomenal zombies would make the same assertions we do about their conscious experiences and about consciousness in general, and with the same conviction, and they would have beliefs on those matters with the same causal and explanatory roles as ours (though, arguably, with different contents). Given this, our claims and beliefs about consciousness afford no evidence for the truth of phenomenal realism, and it is reasonable to regard them as mistaken.

A second argument for illusionism does not depend on the assumption of anti-physicalism. In general, apparent anomalousness is evidence for illusion. If a property resists explanation in physical terms or is detectable only from a certain perspective, then the simplest explanation is that it is illusory. In this light, considerations usually cited in support of a radical approach to consciousness, such as the existence of an explanatory gap, the conceivability of zombies, and the perspectival nature of phenomenal knowledge, afford equal or greater support for illusionism. Given the force of these considerations, if there is even a remote possibility that we are mistaken about the existence of phenomenal consciousness, then there is a strong abductive inference to the conclusion that we are in fact mistaken about it. And there is reason to think that we could be mistaken about it. For our awareness of phenomenal properties would have to be mediated in some way. If the mind is a representational system, then properties must be mentally represented in order to have cognitive, affective, or motivational significance, and phenomenal properties are no exception, regardless of whether they are physical or non-physical. A creature that lacked introspective representations of its phenomenal properties — we might call it a *representational zombie* — would have no cognitive access to its phenomenal properties and would be unable to form beliefs about them, reflect on them, report them, remember them, respond emotionally to them, or act upon them. Its experiences would not be *like anything*, in the second of the senses distinguished earlier.[10] But we have no introspective way of checking the accuracy of our introspective representations, and so cannot rule out the possibility that they are non-veridical. (Indeed, in so far as we can check, through external inspection of our brain states, they appear to be non-veridical; the properties represented do not show up from other perspectives.) For all we know, then, phenomenality might be illusory; and, given its anomalousness, we can abductively infer that it is.

Illusionism has other explanatory advantages too. One is that it permits us to acknowledge both the wonder of phenomenal consciousness and its potency. This is something realists find hard to do. Stressing the magical, non-physical character of phenomenal properties usually means denying them a causal role, while treating them as physical causes means denying that they are as magical as they seem.

---

[10] Compare Rey: 'Postulating qualia properties, whether in the brain or in some special realm, will be of no help unless we have an account of how those properties are assimilated into a person's cognitive life; and it's hard to see how they could be assimilated without being *represented*' (Rey, 2007, pp. 129–30).

But if phenomenal properties are intentional objects, a sort of mental fiction, then we need no longer be embarrassed by them. We can acknowledge how magical and unearthly they are *and* how powerfully they affect us, as intentional objects. In this sense, illusionists may claim to take consciousness more seriously than realists do.

Illusionism also offers an attractive perspective on the function of consciousness. If consciousness has the powerful behavioural influence it seems to have, then we should be able to explain it as an adaptive feature. Again, realists find this hard to do. If consciousness is a matter of pure feel, then it is unclear what function it could perform, and many realists, both radical and conservative, see it as little more than a side effect of perceptual processes. But if consciousness involves an illusion, then new possibilities open. Maybe its function is precisely to give us the impression that we have a magical, non-physical inner life. Humphrey has made a powerful case for such a view (1992; 2006; 2011). He proposes that sensations occur when internalized evaluative responses to stimuli ('sentitions') interact with incoming sensory signals to create complex feedback loops, which, when internally monitored, seem to possess otherworldly, phenomenal properties. This internal 'magic show', Humphrey argues, powerfully affects the creatures that possess it, giving them a new interest in their existence, inducing them to engage more deeply with their environment (onto which they project phenomenal properties), and creating a sense of self, and, in humans, belief in an ego or immaterial soul. These developments, Humphrey argues, were strongly fitness-enhancing, and the magic show has been sculpted by natural selection to promote them. (This is possible since, on Humphrey's view, the mechanisms of sensation are separate from those of perception and can respond to different evolutionary pressures.) Whether or not this account is right (and it has many attractions), it is an excellent illustration of how evolutionary theorizing about consciousness can flourish, once freed from the metaphysical preoccupations of realism.

3. Defending illusionism

This section responds to some common objections to illusionism. It argues that they serve primarily to highlight the commitments of the illusionist approach and that illusionists can accommodate weakened versions of the intuitions on which they draw.

3.1. Denying the data

The most basic objection to illusionism is that it denies the data. To be sure, if all that needed to be explained were the detectable marks of phenomenal consciousness — the related judgments, reports, reactions, dispositions, and so on — then it would be more economical to adopt an illusionist view. But — the objection goes — that is not all that needs to be explained; phenomenal consciousness is itself a datum (Chalmers, 1996, p. 188). Phenomenal properties are not theoretical posits introduced to explain other data, but are themselves core data.

There is a sense in which illusionists can agree. It is a datum that phenomenal properties exist as intentional objects; our introspective reports define a notional introspective world which is as we take it to be. But illusionists do, of course, deny that phenomenal properties exist in the real world, as properties of brain states. We are strongly disposed to think that their existence is an introspective datum, but all observation statements, including ones about our own minds, are open to revision in the light of theory. Our introspective reports are data for a science of consciousness, but they require interpretation and evaluation, and the best explanation for them may be one that denies their reliability (Dennett, 2003; 2007). And, as we have seen, there are strong theoretical reasons to doubt the reliability of our first-person reports about phenomenal consciousness.

If realists are to maintain that phenomenal consciousness is a datum, then they must say that we have a special kind of epistemic access to it, which excludes any possibility of error. And since no causal process could provide such certainty, they must say that this access is not causally mediated. This is indeed what some realists propose. Chalmers holds that we are *directly acquainted* with phenomenal properties (Chalmers, 1996, pp. 192–200). He describes acquaintance as 'a basic sort of epistemic relation between a subject and a property' and says that 'whenever a subject has a phenomenal property, the subject is acquainted with that phenomenal property' (2003, p. 250). Experience is in this sense intrinsically epistemic (1996, p. 196).

This view protects the status of phenomenal consciousness as a datum but does so at a high cost. First, acquaintance can have no psychological significance. In order to talk or think about our phenomenal properties, we need to form mental representations of them, and since representational processes are potentially fallible, the certainty conferred by acquaintance could never be communicated, either to others or even to ourselves, considered as cognitive systems. The price of making consciousness a datum is that the datum is psychologically inert. Second, acquaintance theory assumes that the reactions and associations a sensory episode evokes do not affect its feel, since we are not directly acquainted with them or their effects. Yet there is reason to think that our reactions and associations do shape our sense of what our experiences are like (see Dennett, 1988; 1991, chapter 12; 2005, chapter 4). (It might be replied that these factors influence our *judgments* about the feel of our experiences, not the feels themselves, but this would open a systematic gap between what our experiences are like and what we think they are like — which is, at the least, counter-intuitive.)

Acquaintance theory also comes with heavy metaphysical baggage. It is hard to see how physical properties could directly reveal themselves to us, so the theory plausibly assumes an anti-physicalist view of phenomenal consciousness. Moreover, it may require an anti-physicalist view of the experiencing subject too. If subjects are complex physical organisms, how can they become directly acquainted with phenomenal properties? When cognitive scientists talk of information being *available to the subject*, they mean that it is globally broadcast, available for the flexible control of thought and action, and so on. But events need to be represented in order to be available to the subject in this sense. Talk of acquaintance supposes a non-psychological subject, which exists prior to representational processes, as opposed to being partially constituted by them.

This brings us back to talk of it being *like something* to be us. As noted earlier, such talk may mean simply that we have an introspective awareness of our experiences, generated by representational mechanisms. We might call this *introspective subjectivity*. Illusionists agree that we have introspective subjectivity, though they hold that it is radically misleading. But 'like something' talk can be understood in a stronger sense, as indicating that we possess a subjective dimension that is not the product of introspective mechanisms but arises simply from our being the things we are. Call this *intrinsic subjectivity*. When theorists talk of our being directly acquainted with phenomenal properties, it is intrinsic subjectivity they have in mind; the properties, and our awareness of them, are simply correlates of our physical constitution. Plausibly, then, taking phenomenal consciousness as a datum involves positing intrinsic subjectivity.

Intrinsic subjectivity is, however, deeply mysterious. It is a shadowy companion of physical systems, and we could imagine *any* object possessing it, as panpsychists do. (It might be proposed that only beings with a certain physical structure possess intrinsic subjectivity — perhaps only those that implement an information processing system. But this structure does not *explain* their intrinsic subjectivity, and a creature's reports of what its experiences are like will be the product of introspective mechanisms and will thus manifest introspective subjectivity only.) This does not show that the notion of intrinsic subjectivity is incoherent, but it is, I think, a good reason to explore the idea that it is a fiction created by introspective subjectivity.

3.2. No appearance–reality gap

Another common objection to illusionism is that in the case of qualitative states there is no gap between illusion and reality. Something can look like a Penrose triangle without being a Penrose triangle, but an experience that seems to have a greenish phenomenal character really does have a greenish phenomenal character.[11] As Searle puts it, '*where consciousness is concerned the existence of the appearance is the reality*. If it seems to me exactly as if I am having conscious experiences, then I am having conscious experiences' (Searle, 1997, p. 112, italics in original).[12]

This is often presented as a crushing objection to illusionism, but it is far from compelling. It turns on what we mean by *seeming to have* a greenish experience. If we mean having an introspective experience with the same phenomenal feel as a greenish experience, then, trivially, there is no distinction between seeming and reality. But of course that is not what illusionists mean. They mean introspectively representing oneself as having a greenish experience, and one can do this without having a greenish experience. The objector may reply that, in order to create the illusion of a greenish experience, the introspective representation would have to employ a greenish mode of presentation, which would itself have an introspectable greenish feel. However, illusionists will simply deny this, arguing that the content of introspective representations is determined by non-phenomenal,

---

[11] I follow Levine's practice of using 'greenish' for the (putative) feel associated with perception of a green object (Levine, 2001).

[12] Compare Kripke: 'in the case of mental phenomena there is no "appearance" beyond the mental phenomenon itself' (Kripke, 1980, p. 154).

causal or functional factors.[13] The objector may say that there is a big difference between merely representing oneself as having a greenish experience in such a way and actually having a greenish experience, but that is just the point at issue. The illusionist claims that when we think we are having a greenish experience we are in fact merely misrepresenting ourselves as having one. That claim may be false, but the no-gap objection does not add anything to the case against it. Of course, this requires some account of the content of the representations involved, and providing this will be a major challenge for the illusionist. But it is an independent requirement, and the no-gap objection does not make it harder to meet.

Another version of the no-gap objection might go as follows.[14] It is incoherent to doubt that experiences are as they seem, since experience reports are already reports of how things seem. I may come to doubt my initial claim that there is a green patch in front of me and retreat to the more cautious claim that there seems to be a green patch, but I cannot coherently retreat from that claim to the claim that there *seems to seem* to be a green patch. The first claim expresses all the epistemic caution that is necessary or possible. There is something right about this. We have no everyday procedure for correcting sincere and attentive experience reports, and we treat them as authoritative. But it does not follow that this authority is epistemic. Being cautious about the external world does not make one authoritative about the internal one, and seeming to see a green patch isn't the same as infallibly introspecting a greenish phenomenal feel. Rather, as Dennett suggests, the authority might be more like that which a storyteller has over their fictions (Dennett, 1991, p. 81).[15]

---

[13] Alternatively, illusionists might concede that introspection employs modes of presentation that appear to have phenomenal feels, but argue that this too is an illusion — that introspection misrepresents the modes of presentation as having phenomenal properties they lack. For defence of this option and an argument that it does not generate an infinite regress, see Pereboom (2011, pp. 27–8).

[14] This version was suggested by remarks of Martine Nida-Rümelin, though she might not endorse my presentation of it.

[15] It might be argued that phenomenal properties cannot be illusory, since they serve as sense-data, and it is only when sense-data are interpreted that illusion can arise (Wright, 2008). This is unpersuasive, however, even granting sense-data theory. Introspective representations of phenomenal properties might serve as data in the construction of representations of external reality while themselves misrepresenting internal, neurophysiological reality. (I am grateful to an anonymous referee for raising this objection.)

In a strong form, then, the claim that there is no appearance–reality gap for phenomenal properties is not compelling. A weaker version of the claim is, however, both plausible and compatible with illusionism. From the perspective of a representational theory of mind, the difference between seeming to be aware of a certain phenomenal feel and actually being aware of it is that between having a non-veridical introspective representation of the feel and having a veridical one, and, subjectively, this is no difference at all. In this sense, illusionists can agree that there is no appearance–reality gap for consciousness.

3.3. Who is the audience?

An illusion presupposes an audience. Who is the audience for the illusion of phenomenal consciousness? Illusionists will join Dennett in dismissing the idea that there is an inner arena (a 'Cartesian theatre') where perceptual information is assembled and a phenomenal show presented for an appreciation by an internal observer (Dennett, 1991). But aren't they committed to reinstating a Cartesian theatre as an arena where the *illusion* of phenomenality is presented?

Illusionists may be committed (as many theorists are) to the existence of an inner *representer* of some kind: a system, or set of processes, which generates introspective representations of sensory states. But this need not amount to an observer, still less a conscious one. If we do not need an inner observer to appreciate perceptual representations, why should we need one to appreciate introspective ones? As Dennett argues, once the brain has made a discrimination, there is no need for another brain system to remake it, and all the work of appreciation and reaction can be (and ultimately must be) distributed among numerous unintelligent subsystems (*ibid.*). Similarly, once an introspective representation has been generated, the work of reacting to it — of being impressed by the illusion — can, and must, be distributed across such subsystems. There need be no unified audience for the illusion smaller than the organism as a whole (or at least its central nervous system).

That said, illusionists may posit something like an inner display. Recall Humphrey's proposal that internal feedback loops have been shaped by evolution to create a life-enhancing internal magic show (Humphrey, 2011). Such a show is, however, different from the one in the Cartesian theatre. First, it is not a phenomenal show, though it is represented as one. Second, it is not a redundant re-presentation of information already encoded in the system. The feedback loops are new features, continuously generated, which need to be monitored and represented in order to have psychological effects. Third, the detector system need do no more than generate representations; again, all the work of appreciating and reacting to the show can be parcelled out to subsystems. Finally, (though Humphrey might not agree) the show need not be a single, integrated one, generating a definitive stream of introspective representations. Instead, there might be numerous micropresentations, yielding multiple drafts of sensation (an introspective counterpart of the multiple drafts model of perceptual processing Dennett proposes; Dennett, 1991). Extending the theatrical metaphor, there might be a host of fringe events around the town rather than an official show in a central auditorium.

3.4. Representing phenomenality

Another objection centres on the representation of phenomenality. If there are no phenomenal properties, how do we represent them? How do we acquire phenomenal concepts, and how do these concepts capture the richness of phenomenality? These are central questions for illusionists, and answering them would go a long way towards solving the illusion problem. Here I shall merely make some preliminary remarks and indicate some lines open to the illusionist.

The task of constructing a theory of content for phenomenal concepts is a difficult one, but it is not obvious that it is significantly more difficult for those who hold that these concepts lack referents. Levine questions whether we can explain the richness and determinacy of our phenomenal representations without reference to actual phenomenal properties (Levine, 2001, pp. 146–7). When we think about what an experience is like, he suggests, the phenomenal property itself is included in the thought and serves as its own mode of presentation (*ibid.*, p. 8). The idea that phenomenal concepts quote or incorporate tokens of their referents has been proposed by several theorists (e.g. Chalmers, 2003; Papineau, 2002, pp. 116–25). However, its explanatory power is questionable. Why should incorporating a phenomenal feel into a representational vehicle make the vehicle represent the feel, let alone in a rich and determinate way? (Incorporating iron filings into it wouldn't make it represent iron.) As Rey stresses, some mechanism would be needed to read off features of the incorporated property and represent them to the rest of the system (Rey, 2007, pp. 128–9). But then a secondary, non-quotational account of phenomenal representation would be needed, to which the illusionist could appeal directly.

It is true that illusionism does not sit well with strong externalist views, on which the content of a representation is constituted by causal connections to its referent. Illusionists might argue that phenomenal concepts are compounded from more primitive ones that do refer, or that they have counterfactual causal connections to uninstantiated phenomenal properties. However, there are reasons for finding neither of these options attractive, either for phenomenal concepts or for non-referring concepts generally (Rey, 2005). A better option may be to adopt some form of functional-role semantics for phenomenal concepts, on which their content is fixed by their role in mental processing, including their connections to other concepts, to non-conceptual sensory and introspective representations (their own content determined causally or functionally), and to associations, behavioural dispositions, and so on. (If these functional roles are narrow, 'in the head' ones, the content of our phenomenal representations will be independent of environmental factors — but that is not implausible; see Rey 1998.)

Another possibility is that phenomenal concepts are hybrid ones. Suppose we have a general theoretical concept of a phenomenal property — roughly, that of a simple, intrinsic, immediately known, introspectable property of experience. This concept might be innate, the product of individual theorizing, or culturally acquired. Suppose, too, that we have capacities to introspectively recognize different types of sensory states when they occur, and associated recognitional concepts for the states identified. Then phenomenal concepts might be hybrid ones combining the general theoretical concept with specific recognitional ones. For example, the concept of a certain shade of phenomenal red might be that of *this kind of phenomenal property*, where 'this kind' refers to the kind picked out by the recognitional capacity exercised while having an experience of the relevant type. Of course, if illusionism is true, that capacity does not pick out a phenomenal property; it picks out a complex physical one. So the hybrid concept fails to refer. (Compare 'that kind of ectoplasm' said by a credulous spectator at a séance.) Indeed, the theoretical concept may inform our introspective awareness, so that we mistakenly introspect sensory states *as* phenomenal, just as we might misperceive a flat hologram *as* a three-dimensional object (perhaps even an impossible one, such as a Penrose triangle). A hybrid theory like this may be able to account for many of our intuitions about phenomenal consciousness, rendering illusionism more palatable. If introspection employs recognitional concepts, it may present its objects as being simple, ineffable, and immediately known, but if it is also theoretically informed, it may at the same time radically misrepresent them.

4. Facing up to the illusion problem

Illusionism replaces the hard problem with the illusion problem — the problem of explaining how the illusion of phenomenality arises and why it is so powerful. This problem is not easy but not impossibly hard either. The method is to form hypotheses about the underlying cognitive mechanisms and their bases in neurophysiology and neuroanatomy, drawing on evidence from across the cognitive sciences. There are many theoretical options available, and I have indicated some dimensions along which illusionist theories may differ. Some of the issues and positions will be similar to those discussed by conservative realists, but they will assume a new aspect once the commitment to realism is dropped, and we can expect new connections to appear and new theoretical options to present themselves.

Most people find it incredible, even ludicrous, to suppose that phenomenal consciousness is illusory. But if the illusion has been hardwired into our psychology for good evolutionary reasons, then that is to be expected. The question is not whether illusionism is intuitively plausible, but whether it is rationally compelling. If we had a detailed and well-supported illusionist theory, which fully explained our reports, judgments, and intuitions about our own consciousness, would we still want to insist, on reflection, that a hard problem remained? The best way to find out will be to try to construct such a theory.

Our introspective world certainly seems to be painted with rich and potent qualitative properties. But, to adapt James Randi, if Mother Nature is creating that impression by actually equipping our experiences with such properties, then she's doing it the hard way.[16]

---

[16] Earlier versions of this article were presented at The Open University and the University of Crete, and at a 'consciousness cruise' organized by Dmitry Volkoff and the Moscow Center for Consciousness Studies in June 2014, where Jesse Prinz presented a comment on it. My thanks to Jesse and to the audiences on those occasions, mentioning in particular Philip Goff, Martine Nida-Rümelin, Carolyn Price, and Michael Tye. Thanks are also due to Ned Block, Daniel Dennett, Eileen Frankish, Nicholas Humphrey, and Maria Kasmirli for their advice and suggestions. I am especially grateful to David

References

Blackmore, S. (2011) *Zen and the Art of Consciousness*, London: Oneworld Publications.
Carruthers, P. (2000) *Phenomenal Consciousness: A Naturalistic Theory*, Cambridge: Cambridge University Press.
Chalmers, D.J. (1996) *The Conscious Mind: In Search of a Fundamental Theory*, Oxford: Oxford University Press.
Chalmers, D.J. (2003) The content and epistemology of phenomenal belief, in Smith, Q. & Jokic, A. (eds.) *Consciousness: New Philosophical Perspectives*, pp. 220–272, Oxford: Oxford University Press.
Dennett, D.C. (1988) Quining qualia, in Marcel, A.J. & Bisiach, E. (eds.) *Consciousness in Modern Science*, pp. 42–77, Oxford: Oxford University Press.
Dennett, D.C. (1991) *Consciousness Explained*, New York: Little, Brown.
Dennett, D.C. (2003) Who's on first? Heterophenomenology explained, *Journal of Consciousness Studies*, **10** (9–10), pp. 19–30.
Dennett, D.C. (2005) *Sweet Dreams: Philosophical Obstacles to a Science of Consciousness*, Cambridge, MA: MIT Press.
Dennett, D.C. (2007) Heterophenomenology reconsidered, *Phenomenology and the Cognitive Sciences*, **6** (1–2), pp. 247–270.
Frankish, K. (2012) Quining diet qualia, *Consciousness and Cognition*, **21** (2), pp. 667–676.
Hall, R.J. (2007) Phenomenal properties as dummy properties, *Philosophical Studies*, **135** (2), pp. 199–223.
Harman, G. (1990) The intrinsic quality of experience, in Tomberlin, J. (ed.) *Philosophical Perspectives, Vol. 4, Action Theory and Philosophy of Mind*, pp. 31–52, Atascadero, CA: Ridgeview.
Hill, C.S. (1997) Imaginability, conceivability, possibility and the mind–body problem, *Philosophical Studies*, **87** (1), pp. 61–85.
Hill, C.S. & McLaughlin, B.P. (1999) There are fewer things in reality than are dreamt of in Chalmers's philosophy, *Philosophy and Phenomenological Research*, **59** (2), pp. 445–454.
Howell, R. (2015) The Russellian monist's problems with mental causation, *The Philosophical Quarterly*, **65** (258), pp. 22–39.
Humphrey, N. (1992) *A History of the Mind: Evolution and the Birth of Consciousness*, New York: Simon and Schuster.
Humphrey, N. (2006) *Seeing Red: A Study in Consciousness*, Cambridge, MA: Harvard University Press.
Humphrey, N. (2011) *Soul Dust: The Magic of Consciousness*, Princeton, NJ: Princeton University Press.
Kripke, S.A. (1980) *Naming and Necessity*, new ed., Oxford: Blackwell.
Levine, J. (2001) *Purple Haze: The Puzzle of Consciousness*, Oxford: Oxford University Press.
Loar, B. (1990) Phenomenal states, in Tomberlin, J.E. (ed.) *Philosophical Perspectives, Vol. 4, Action Theory and Philosophy of Mind*, pp. 81–108, Atascadero, CA: Ridgeview.

Chalmers for his detailed comments on earlier drafts, from which the article has benefited considerably.

Noë, A. (ed.) (2002) *Is the Visual World a Grand Illusion?*, Exeter: Imprint Academic.
Papineau, D. (2002) *Thinking about Consciousness*, Oxford: Oxford University Press.
Pereboom, D. (2011) *Consciousness and the Prospects of Physicalism*, New York: Oxford University Press.
Place, U.T. (1956) Is consciousness a brain process?, *British Journal of Psychology*, **47**, pp. 44–50.
Prinz, J.J. (2004) The fractionation of introspection, *Journal of Consciousness Studies*, **11** (7–8), pp. 40–57.
Randi, J. (1997) Science and pseudoscience, in Terzian, Y. & Bilson E. (eds.) *Carl Sagan's Universe*, pp. 170–108, Cambridge: Cambridge University Press.
Rey, G. (1992) Sensational sentences switched, *Philosophical Studies*, **68** (3), pp. 289–319.
Rey, G. (1995) Towards a projectivist account of conscious experience, in Metzinger, T. (ed.) *Conscious Experience*, pp. 123–142, Exeter: Imprint Academic.
Rey, G. (1998) A narrow representationalist account of qualitative experience, in Tomberlin, J.E. (ed.) *Philosophical Perspectives, Vol. 12, Language, Mind, and Ontology*, pp. 435–457, Boston, MA: Blackwell.
Rey, G. (2005) Philosophical analysis as cognitive psychology: The case of empty concepts, in Cohen, H. & Lefebvre, C. (eds.) *Handbook of Categorization in Cognitive Science*, pp. 71–89, Amsterdam: Elsevier.
Rey, G. (2007) Phenomenal content and the richness and determinacy of colour experience, *Journal of Consciousness Studies*, **14** (9), pp. 112–131.
Searle, J.R. (1997) *The Mystery of Consciousness*, New York: The New York Review of Books.
Smart, J.J.C. (1959) Sensations and brain processes, *The Philosophical Review*, **68** (2), pp. 141–156.
Strawson, G. (2006) Realistic monism: Why physicalism entails panpsychism, in Freeman, A. (ed.) *Consciousness and its Place in Nature*, pp. 3–31, Exeter: Imprint Academic.
Tartaglia, J. (2013) Conceptualizing physical consciousness, *Philosophical Psychology*, **26** (6), pp. 817–838.
Tye, M. (1995) *Ten Problems of Consciousness: A Representational Theory of the Phenomenal Mind*, Cambridge, MA: MIT Press.
Tye, M. (2000) *Consciousness, Color, and Content*, Cambridge, MA: MIT Press.
Wright, E. (2008) Introduction, in Wright, E. (ed.) *The Case for Qualia*, pp. 1–42, Cambridge, MA: MIT Press.

Katalin Balog
Illusionism's Discontent

Abstract: *Frankish positions his view, illusionism (a.k.a. eliminativist physicalism), in opposition to what he calls radical realism (dualism and neutral monism) and conservative realism (a.k.a. non-eliminativist physicalism). Against radical realism, he upholds physicalism. But he goes along with key premises of the* gap arguments *for radical realism, namely, 1) that epistemic/explanatory gaps exist between the physical and the phenomenal, and 2) that every truth should be perspicuously explicable from the fundamental truth about the world; and he concludes that because physicalism is true, there could not be phenomenal truths. I think he is wrong to accept 2; and even if he was right to accept it, the more plausible response would be not to deny the existence of qualia but to deny physicalism. In either case, denying the existence of qualia is the wrong answer. I present a physicalist realist alternative that refutes premise 2 of the gap argument; I also make a general case against the scientism that accompanies Frankish's metaphysics.*

Keith Frankish has spelt out an interesting case for illusionism by positioning it as the least unpalatable of three rather defective positions on consciousness. The other two are what he calls *radical realism* (which covers dualism and neutral monism) and *conservative realism*, a.k.a. non-eliminativist physicalism. His view, which can be described as *conservative eliminativism*, denies the radical realist's contention that the world, and especially those portions of it that have minds around, consists of more than just physical stuff; but also takes issue with conservative realism (non-eliminativist physicalism) despite sharing its commitment to a purely physical ontology. Against conservative realism, he argues that it is not possible to be a realist about consciousness and still remain a physicalist; and he thinks our reasons for physicalism override our reasons to believe in qualia. My

Correspondence:
Email: kbalog@andromeda.rutgers.edu

own perspective is just the reverse. I hold that our reasons to believe in qualia are stronger than our reasons for physicalism — but I also think that qualia realism is compatible with physicalism.

My remarks will be mostly critical; but I want to register that I have found Frankish's views not only very engaging, but also extraordinarily helpful in orienting myself in the philosophical landscape. In the first section I will look at Frankish's general arguments for illusionism and what a conservative realist (i.e. non-eliminativist physicalist) can say in defence of realism. In the second part I will look at the plausibility of illusionism in its own right, and in the third highlight some specific problems with his account of the nature of the 'illusion'.

1. Physialism vs. scientism

The most compelling consideration Frankish presents for illusionism is related to a well-known family of arguments, let's call them *gap arguments*, that aim to refute physicalism by appeal to various (epistemic, conceptual, and explanatory) gaps between physical and phenomenal descriptions of the world which, according to these arguments, provide *a priori* reason to reject physicalism. Frankish puts his concern with conservative (physicalist) realism in a very similar vein:

> The central problem, of course, is that phenomenal properties seem too weird to yield to physical explanation. They resist functional analysis and float free of whatever physical mechanisms are posited to explain them. (Frankish, this issue, p. 25)

Here is how this leads to illusionism:

> In general, apparent anomalousness is evidence for illusion. If a property resists explanation in physical terms or is detectable only from a certain perspective, then the simplest explanation is that it is illusory. In this light, considerations usually cited in support of a radical approach to consciousness, such as the existence of an explanatory gap, the conceivability of zombies, and the perspectival nature of phenomenal knowledge, afford equal or greater support for illusionism. (pp. 27–8)

After aligning himself with the gap arguments for illusionist conclusions, Frankish continues to hammer away at realism. He thinks our introspective judgments about conscious experience should be discounted because

> ...through external inspection of our brain states, they appear to be non-veridical; the properties represented do not show up from other perspectives. (p. 28)

These last remarks expose an important aspect of Frankish's illusionism. He combines the crucial premises of the gap argument — in this case, that all truths should be perspicuously explicable from the fundamental truths, and the observation that there are no physical explanations of qualia — with physicalism to argue that qualia don't exist. This combination of views supports and feeds on his *scientism*: that the best way to study everything is through science. Though scientism goes beyond the metaphysical position Frankish explicitly argues for, it is a plausible extension of it and is certainly in the background of the views of other notable illusionists Rey (1995) and Dennett (1991).

I think scientism is, for all the wonderful progress science has made, a theoretical mistake; but one with adverse practical consequences. I find the denial of qualia utterly implausible, and scientism a misguided way to approach one's life. There is a concern, expressed in Western philosophy most forcefully by Kierkegaard, namely that our experience of life matters in ineffable ways that no objective understanding of the world can capture. Wittgenstein, in a well-known letter to Ludwig von Ficker, the publisher of the *Tractatus*, claimed that 'the whole point of the book is to show that what is important lies in what cannot be expressed' in a scientific language (Wittgenstein, 1979, p. 94).

Suppose there was a super-intelligent organism — in a twist on Frank Jackson's (1982) knowledge argument — that lacked any feeling or experience, a creature of pure thought. A purely scientific account of humans — though very far from reality — is perhaps not an impossibility. So such a being could know everything about humans in biological, neuroscientific, and information processing terms — even though she lacked the introspective understanding normal humans have of their subjective reality. Such a creature would arguably know nothing of value, meaning, and human significance.

This is, of course, a crude opposition. We hardly ever relate to anything purely objectively or purely subjectively. But as Kierkegaard pointed out, modern life supports a tendency toward objectivity, while, in fact, one needs to become subjective, fully immersed in one's consciousness, to properly understand oneself. He said:

> ...this is the wonder of life, that each man who is mindful of himself knows what no science knows, since he knows who himself is...
> (Kierkegaard, 1980, pp. 78–9)

So I reject scientism in any of its forms. I believe that if the existence of qualia was incompatible with physicalism it would be reason to reject physicalism (rather than embrace it and scientism, as Frankish does). But I do not think that the existence of qualia and subjectivity is incompatible with physicalism. While I would rather give up physicalism before I'd embrace scientism and give up belief in first-person authority about the mind, I do not think this is a choice forced on physicalism. One can have physicalism without the scientism and illusionism about qualia. It is tenable that purely physical minds conceive of their own contents, and especially their (real) qualitative states in a way that is inaccessible to and isolated from the point of view of science. Of course, given physicalism, no facts exist that do not have an exhaustive third-person account, but this doesn't exclude first-person takes on the world and the mind that cannot be — perspicuously — explained from the scientific perspective.

This is exactly the perspective of what has become known as the 'phenomenal concept strategy' (Stoljar, 2005). The strategy is based on an idea first articulated by Brian Loar (1990; 1997) that the epistemic, conceptual, and explanatory gaps between phenomenal and physical descriptions can be explained by appeal to the nature of phenomenal *concepts*, thereby obviating both the illusionist and the anti-physicalist response to the gaps. Phenomenal concepts, on this proposal, involve unique cognitive mechanisms, but none that could not be fully physically implemented.

The key idea of the phenomenal concept strategy is to give an account of how phenomenal concepts can refer to conscious states *directly* and yet in a *substantive* manner, even while supposing that they refer to physical (plausibly, neural) states in the brain, via entirely physical mechanisms. On this view, both qualia and the phenomenal concepts we apply to them are physical; but phenomenal concepts involve unique cognitive mechanisms that set them apart — in fact, isolate them — conceptually from scientific concepts. Loar's core idea is that when a person is having a particular experience she can deploy a concept that refers directly to the experience and that in some way involves in its mode of presentation the very experience it refers to, and that this account of phenomenal concepts is entirely neutral with regard to the metaphysical status of conscious states; that is, entirely neutral on the question of whether qualia are physical or irreducibly mental. It also explains why physicalism about qualia *seems* to be puzzling.

One way to understand this idea, the one I favour, is to hold that phenomenal concepts are partly constituted by tokens of the phenomenal experiences they refer to (Balog, 2012a,b; Block, 2006;

Chalmers, 2003; and Papineau, 2002). On this view, a token phenomenal experience is *part* of the token concept referring to it, and the experience — at least partly — determines that the concept refers to the experience it contains a token of. Of course, 'part' does not mean 'spatial part' but rather that it is *metaphysically* impossible to token the concept without tokening an instance of its referent.

This account of phenomenal concepts is not intended to apply to all concepts that refer to phenomenal states but only to what we might call 'direct phenomenal concepts'. Of course most of our reference to phenomenal states and qualia do not contain the phenomenal states themselves. Clearly, a person can token a concept that refers to pain without her literally experiencing pain — these can be called 'indirect phenomenal concepts' — as when she replies to her dentist's question with 'I am not in pain' or when one sees another person stub her toe and thinks 'that hurts'. But for the purposes of discussion it is appropriate to focus on direct phenomenal concepts since these are the ones that generate the puzzlement over qualia.

If the above account is right, phenomenal concepts have very special *realization states*: the neural states realizing these concepts are instances of the very same neural states types the concepts refer to. What is so special about phenomenal concepts, on this account, is not only that their *realization states* are instances of their *referents*, but that this very fact is crucially involved in determining their meaning. In other words, not only are these concepts realized by instances of their referents, but they refer to what they do at least in part *in virtue* of this fact. This is, of course, very different from any other concept. Most concepts are not realized by tokens of their referents at all; but even those — like the concept ATOM — that are, mean what they do completely independently of this fact about realization. This also means that the cognitive mechanisms involved in phenomenal concepts guarantee that we will be puzzled by how qualia fits in with the brain, whether or not physicalism is true.

Let me briefly explain how the constitutional theory of phenomenal concepts accounts for the explanatory gap. Recall that the problem of the explanatory gap is that no amount of knowledge about the physical facts (brain functioning and so on) is able to explain why a particular brain state/process has a particular feel, e.g. feels giddy. This contrasts, for example, with the way the fact that water is composed of H_2O molecules together with physical and chemical laws explains why water is potable, transparent, and so on. The explanation of why H_2O behaves in watery ways (together with the fact that water is composed of H_2O molecules) straightforwardly explains the behaviour of water. Since we can't explain why a brain state feels giddy in neurophysiological terms, we can't close the physical–phenomenal explanatory gap.

The constitutional account explains the gap by appealing to the *substantial* and *direct* grasp phenomenal concepts afford of their referent. When I focus on the phenomenal state, I have a 'substantial' grasp of its nature. I grasp what it is like to be in that phenomenal state — in terms of *what it's like to be in that same state*. This is what the constitutional account captures. And because this grasp is at the same time *direct*, that is, independent of any causal or functional information (unlike in the case of WATER), information about the functioning of the brain simply won't explain *what it's like to be* in that state.

What exactly is this *substantial* insight into the nature of phenomenal states? If phenomenal concepts are partly constituted by phenomenal states, our knowledge of the presence of these states (when we apply these direct phenomenal concepts) is not mediated by something distinct from these states. Rather the state itself serves as its own mode of presentation. When I focus on the phenomenal quality of an experience — not on what it represents but on its qualitative character — my representation contains that very experience. Thinking about it and simply having the experience will then share something very substantial, very spectacular: namely the phenomenal character of the experience. Being aware of our phenomenal states — being *acquainted* with them (Russell, 1910) — is the special, intimate epistemic relation we have to our phenomenal experience through the *shared phenomenality* of experience and thought. Shared phenomenality produces the sense that one has a direct insight into the nature of the experience. And it seems puzzling, to say the least, how this nature could be physical. But it is important to notice that this kind of direct insight (via shared phenomenality of thought and experience) into the nature of conscious experience does not reveal anything about the metaphysical nature of phenomenality. It is not the same sense of 'insight into the nature of X' as a scientific analysis of a brain state would provide. The one involves *having* the state, the other, analysing it into its components, which are very different activities.

So the constitutional account of phenomenal concepts offers a solution to the mind–body problem that steers clear of both radical realism and illusionism. As a matter of fact, it tackles head on the main reason Frankish cites for illusionism, that is, the existence of epistemic/conceptual/explanatory gaps between the physical and the phenomenal; and concludes that it is not a good reason to give up either physicalism or realism. Though Frankish mentions this approach, he dismisses it quickly, without much discussion.

In contrast, I think we should not take qualia lightly, and should look very seriously at views that could ground a physicalist realism about qualia. Illusionism should only be considered after all other avenues have been exhausted. Just as illusionism about the external world is hard to take seriously even though in a certain sense it fits our data well, illusionism about qualia should not be invoked lightly. Belief in the existence of qualia is just as foundational for our worldview as — and some would say even more so than — belief in the existence of the external world. It takes much more to make it a plausible position than simply showing that it is — at least *prima facie* — coherent and fits some other, initially plausible, principles.

2. The plausibility of illusionism

Frankish makes an attempt to neutralize the inherent implausibility of illusionism by explaining our stubborn (and supposedly erroneous) sense that we are phenomenally conscious. According to illusionism, when we are introspectively aware of our sensory states our awareness is partial and distorted, leading us to *misrepresent* the states as having phenomenal properties. So though nothing in this world, as a matter of fact, instantiates phenomenal properties, it still *appears* to us that our experiences do.

Frankish then tries to coax us to see his claim that introspection misrepresents as plausible. But the analogies he supplies fail to convince. He refers to Dennett's analogy with computer graphics.

> The icons, pointers, files, and locations displayed on a computer screen correspond in only an abstract, metaphorical way to structures within the machine, but by manipulating them in intuitive ways we can control the machine effectively, without any deeper understanding of its workings. The items that populate our introspective world have a similar status, Dennett suggests. They are metaphorical representations of real neural events, which facilitate certain kinds of mental self-manipulation but yield no deep insight into the processes involved. (Frankish, this issue, p. 16)

As Frankish himself points out, this analogy is quite imperfect. Introspection is not like looking at a computer screen, and computer icons are not misrepresentations either. No one who uses a computer really thinks that computer files are located on the screen or that they look like their icons.

Frankish also cites Rey (1995, pp. 137–9), who explains the illusion of qualia as being similar to other illusions where stabilities in our reactions to the world induce us to project corresponding properties onto the world (e.g. our stable personal concerns and reactions to others lead us to posit stable, persisting selves as their objects). Again, while this mechanism is contentious even as an explanation of our concept of self, it doesn't seem to be similar at all to how our introspective phenomenal concepts work. When we form an introspective phenomenal concept of a pain sensation in the act of attending to it, we are not conceptually engaging, much less projecting stable personal concerns and reactions; we simply mentally note the pain. We can be aware of qualia via simple direction of attention. Everybody can do this and there is nothing tendentious about it. It is just a bedrock feature of what it is to be a human being.

Despite Frankish's examples and explanations, I find illusionism extraordinarily implausible simply because it flies in the face of one of the most fundamental ways the world presents itself to us: the awareness of our own mind. Illusionism perhaps sounds plausible, or at least conceivable, from the third-person, scientific perspective we can take on mental representation. From this point of view, it is possible to argue that organisms have no introspective way of checking the accuracy of their introspective representations, and so they cannot rule out the possibility that these representations are non-veridical.

It is clearly the case that science and objective philosophizing might dislodge deeply held common sense views. Obvious examples are the nature of physical objects, and, more controversially, the nature of the self and free will. But the case of qualia is not like that. Arguably, *pace* Frankish, there are no scientific or philosophical discoveries that force us to give up belief in qualia; and there is no demonstrable conceptual incoherence in our introspective concepts of qualia. So the question comes down to the epistemic authority accorded to introspective awareness vs. scientific theorizing.

It seems that Frankish has a negative view of qualia and their role in our life. He, in the strange expression he uses, finds qualia potentially *embarrassing*. He thinks that illusionism can eliminate the embarrassment and clear away the obstacles from taking qualia seriously. As he puts it:

> But if phenomenal properties are intentional objects, a sort of mental fiction, then we need no longer be embarrassed by them. We can

acknowledge how magical and unearthly they are *and* how powerfully they affect us, as intentional objects. In this sense, illusionists may claim to take consciousness more seriously than realists do. (p. 29)

But there are also signs that Frankish is not completely at ease with illusionism.

3. An illusion of illusion?

Some aspects of Frankish's presentation of illusionism strike me as covert attempts to smuggle qualia in through the back door. He seems to me to appeal — illicitly — to qualitative properties in explicating and motivating his own denial of them. First I will talk about problems regarding reference to non-existent qualia, then I will make some remarks about Frankish's treatment of what it is like to have an experience.

3.1. Phenomenal concepts

The heart of illusionism is the view that introspection misrepresents sensory experience as having certain qualitative properties nothing in fact has. But given how vivid our grasp of these allegedly uninstantiated properties are, one is owed an explanation how, and through what mechanism, we can latch onto something that doesn't exist in such a revealing way. The story, of course, cannot run along the same lines as the story for our concept 'unicorn' does; our phenomenal concepts are simple and direct in a way that precludes construction from other, *bona fide* referring concepts. As Levine (2001, pp. 146–7) has observed, there appears to be a problem accounting for the infinitely rich ways in which these concepts apparently refer to an infinitely rich field of phenomenal properties. It is very challenging to explain what it means to represent phenomenality directly — if there is no such thing.

Frankish's answer doesn't come close to meeting the challenge:

> A better option may be to adopt some form of functional-role semantics for phenomenal concepts, on which their content is fixed by their role in mental processing, including their connections to other concepts, to non-conceptual sensory and introspective representations (their own content determined causally or functionally), and to associations, behavioural dispositions, and so on. (p. 36)

This is, unfortunately, little more than hand-waving about how reference to non-existent (or non-instantiated) properties with direct modes of presentation can be established. And in fact there is reason to be suspicious that such an account could ever be found. The problem can be stated as a dilemma. Either introspective concepts refer to real properties so introspection results in meaningful even though erroneous representations, or they don't really refer to any property. In the first case, one just wonders what miracle could ensure that people refer directly to all those wonderful qualitative and subjective properties even though nothing in the world instantiates them? And in the second, all our introspective qualia representations would simply be meaningless, mental junk, so to speak. So the account either requires a miracle, or collapses into meaninglessness.

I suspect that there is a tacit appeal to qualia in illusionism which makes it initially plausible. Because in reality we are all acquainted with qualia, we don't get worried about the idea that introspective representations can refer to them. But when we realize what the account says, namely that nothing has qualia, it should really strike us as utterly miraculous that, *if the account was true*, we could refer to them.

3.2. The 'what it is like' of experience

While illusionism denies the existence of qualia, Frankish seems to want to allow that there is something it is like to think about experience, and even talks about 'introspective subjectivity'. One might wonder: where is the illusion then? It would be pointless to deny that experience has qualitative, subjective properties only to allow introspective representation of experience to have them. That would still be a realist position. So the illusionist's 'what it is like' must be construed otherwise than as 'having qualitative features'. Here is how Frankish explains the distinction:

> Illusionists can say that one's experiences are like something if one is aware of them in a functional sense, courtesy of introspective representational mechanisms. Indeed, this is a plausible reading of the phrase; experiences are like something for a creature, just as external objects are like something for it, if it mentally represents them to itself. Illusionists agree that experiences are like something in this sense, though they add that the representations are non-veridical, misrepresenting experiences as having phenomenal properties (what-it's-likeness in the first sense). (p. 23)

Even assuming that it is supposed to be a constitutive account of what-it's-like-ness, this is not very helpful. To create a new sense of 'what it is like', it not only has to be different from 'having qualia properties', but it also has to be discernably different from concepts of mere function and representation — otherwise invoking the expression 'what it is like' is just a funny way to dress up 'function' and 'representation' talk. It is a redefinition of the concept 'what it is like', rather than a new understanding of it. It introduces no new insight about experience.

Nevertheless, Frankish seems to think that his new concept of 'what it is like' really does speak to our ordinary notion of what it is like. He uses the account to dispel misconceptions about zombies:

> [The illusion] depends on a complex array of introspectable sensory states, which trigger a host of cognitive, motivational, and affective reactions. If we knew everything about these states, their effects, and our introspective access to them, then, illusionists say, we could not clearly imagine a creature possessing them without having an inner life like ours. (p. 23)

Assuming that this is not merely a claim about imagination, but about conceivability, this indicates that, according to Frankish, suitable claims about representation and function conceptually necessitate claims about what it is like to be a creature entertaining those representations. This would indicate that he indeed provided a functional-representational notion of what-it's-like-ness — not merely a pseudo-what-it's-like-ness concept. But this flies in the face not only of what most philosophers believe about phenomenal concepts (that they do not have conceptually sufficient conditions in functional/representational terms), but also of the main reason Frankish presented for illusionism: the gap arguments. If zombies are unimaginable, indeed inconceivable, then the case for radical realism vanishes and conservative realism becomes a viable option.

So I think Frankish overplays his hand with his claim that he can account for the what-it's-like-ness of experience. It might be that, even for an illusionist, the allure of qualia is too strong to resist. But trying to have his cake and eat it will in philosophy, as in the kitchen, get you into trouble when members of your family arrive.

References

Balog, K. (2012a) Acquaintance and the mind–body problem, in Hill C. & Gozzano, S. (eds.) *The Mental, the Physical*, pp. 16–43, Cambridge: Cambridge University Press.

Balog, K. (2012b) In defense of the phenomenal concept strategy, *Philosophy and Phenomenological Research*, **84** (1), pp. 1–23.

Block, N. (2006) Max Black's objection to mind–body identity, in Zimmerman, D. (ed.) *Oxford Studies in Metaphysics*, II, pp. 3–78, Oxford: Oxford University Press.

Chalmers, D. (2003) The content and epistemology of phenomenal belief, in Smith, Q. & Jokic, A. (eds.) *Consciousness: New Philosophical Perspectives*, pp. 220–273, Oxford: Oxford University Press.

Dennett, D.C. (1991) *Consciousness Explained*, New York: Little, Brown.

Frankish, K. (this issue) Illusionism as a theory of consciousness, *Journal of Consciousness Studies*, **23** (11–12).

Jackson, F. (1982) Epiphenomenal qualia, *Philosophical Quarterly*, **32**, pp. 127–136.

Kierkegaard, S. (1980) *The Concept of Anxiety*, Thomte, R. (trans. & ed.), Princeton, NJ: Princeton University Press.

Levine, J. (2001) *Purple Haze: The Puzzle of Consciousness*, Oxford: Oxford University Press.

Loar, B. (1990) Phenomenal states, *Philosophical Perspectives*, **4**, pp. 81–108.

Loar, B. (1997) Phenomenal states: Second version, in Block N., Flanagan, O. & Guzeldere, G. (eds.) *The Nature of Consciousness: Philosophical Debates*, pp. 597–616, Cambridge, MA: MIT Press.

Papineau, D. (2002) *Thinking About Consciousness*, New York: Oxford University Press.

Rey, G. (1995) Towards a projectivist account of conscious experience, in Metzinger, T. (ed.) *Conscious Experience*, pp. 123–142, Exeter: Imprint Academic.

Russell, B. (1910) Knowledge by acquaintance and knowledge by description, *Proceedings of the Aristotelian Society*, **11**, pp. 108–128. Reprinted in Russell, B. (1963) *Mysticism and Logic*, pp. 152–167, London: Allen and Unwin.

Stoljar, D. (2005) Physicalism and phenomenal concepts, *Mind and Language*, **20** (2), pp. 296–302.

Wittgenstein, L. (1979) Letters to Ludwig Ficker, in Luckhardt, C.G. (ed.) *Wittgenstein: Sources and Perspectives*, Gilette, B. (trans.), Ithaca, NY: Cornell University Press.

Susan Blackmore

Delusions of Consciousness

Abstract: *Frankish's illusionism aims to replace the hard problem with the illusion problem; to explain why phenomenal consciousness seems to exist and why the illusion is so powerful. My aim, though broadly illusionist, is to explain why many other false assumptions, or delusions, are so powerful. One reason is a simple mistake in introspection. Asking, 'Am I conscious now?' or 'What is consciousness?' makes us briefly conscious in a new way. The delusion is to conclude that consciousness is always like this instead of asking, 'What is it like when I am not asking what it is like?' Neuroscience and disciplined introspection give the same answer: there are multiple parallel processes with no clear distinction between conscious and unconscious ones. Consciousness is an attribution we make, not a property of only some special events or processes. Notions of the stream, contents, continuity, and function of consciousness are all misguided as is the search for the NCCs.*

In his clear and helpful survey of the illusionist position, Frankish convincingly argues for taking illusionism seriously and for replacing the hard problem with the illusion problem. This is timely and welcome. From teaching consciousness courses to undergraduates for many years I know how hard it is to get students to see round their strongly held intuitions about dualism, zombies, the knowledge argument, and so on. Illusionism cuts through these intuitions and redirects the problem towards asking how and why we have these intuitions and why we seem to be conscious in the way we do. Yet I do not see that any illusionist theory has entirely done away with the hard problem and in the mean time I am being cautious.

Correspondence:
Email: sue@susanblackmore.uk

The illusionist research programme is clearly worth pursuing — indeed it may be the only research programme worth pursuing. Yet if I understand Frankish correctly, I am not quite an illusionist in his sense. Frankish describes 'the basic illusionist claim that introspection delivers a partial, distorted of view of our experiences, misrepresenting complex physical features as simple phenomenal ones' (p. 18). I agree that introspection provides a distorted view but not that it necessarily ends up with 'simple phenomenal ones'. Introspection may sometimes do this and sometimes do the opposite as, for example, in change blindness. This reveals that people believe their own visual worlds contain far more information than they do. This is why I titled the original paper on change blindness, 'Is the Richness of Our Visual World an Illusion?' (Blackmore *et al.*, 1995). Subsequent studies on change blindness blindness showed that people really do believe they have more information available than they have (Levin *et al.*, 2000). In other words they are convinced that they are experiencing a very complex and detailed inner visual world when in fact they are relying on their ability to look again (Simons and Rensink, 2005).

For this and other reasons I refer to consciousness as illusory and have tried to work out how and why the illusions arise. Yet my enterprise is far less ambitious than Frankish's. I am less concerned with explaining 'why experiences seem to have phenomenal properties' (his 'illusion problem') and more concerned with why they lead to false theorizing. Many researchers seem, despite their fervent denials, to be Cartesian materialists (Dennett, 1991) and mired in lurking dualist assumptions. This hampers research by leading them to ask the wrong questions and look in the wrong places for answers. So my aim is not the bold one of escaping the hard problem but the more limited one of exposing and understanding the power of these false ideas. Because this theorizing is largely intellectual, the assumptions might be called 'delusions' rather than 'illusions'.

1. Delusionism

Sitting here at my desk I can feel the keys beneath my cold fingers, hear birds singing outside the window, and see a wealth of colour in the trees and flowers and sky. When introspecting on this it is all too easy to leap from immediate sensations to theorizing about what 'must' be going on. Failing to stop at sensations of cold or dappled green or bird song, we readily assume a 'me' who is the subject of a stream of conscious experiences, and that these arise from (or emerge from, are produced by, or even 'are') a few brain processes while everything else remains 'unconscious'. Yet all this is fantasy. It is these proto-theories, not the cold, the green, or the singing, that constitute the delusion.

Frankish asks why the illusion of phenomenality is so powerful. I would ask the related, but different, question of why this deluded theorizing is so tempting and so powerful. The answer, I suggest, is amusingly simple, if counter-intuitive. Ask yourself this question:

'Am I conscious now?'

I guess that your answer is 'yes'. It is very hard, though not impossible, to answer truthfully 'no' (more of this later). So, whenever we ask this question, we reply, 'Yes — I am conscious now'. We can ask many times and always get the same answer, making it easy to conclude that life is always like this, that I am continuously present and experiencing a stream of thoughts and perceptions. So we assume a continuity and unity which is simply not true. The illusion is powerful because it is so hard to answer a different question — what is it like the rest of the time? What is it like when I am not asking what it is like?

This illusion is similar in structure to why 'the richness of our visual world is an illusion'. The reason we seem to see a richly detailed and picture-like view of the world is that wherever we look we see rich detail. We can always look again and the detail appears 'just in time', so it *seems* that the whole picture is in our minds rather than where it always was, out in the world (O'Regan and Noë, 2001). This illusion is powerful because it is hard to answer the question, 'What does something look like when I am not looking at it?'

There is thus something very curious about the nature of consciousness — that looking into consciousness reveals only what it is like when we are looking into it — and most of the time we are not. So introspecting on our own minds is thwarted by the very fact of introspecting.

William James had a wonderful metaphor for this more than a century ago. Trying to observe the 'flights' as well as the 'perchings' in his 'stream of consciousness', he said 'The attempt at introspective analysis in these cases is in fact like seizing a spinning top to catch its motion, or trying to turn up the gas quickly enough to see how the darkness looks' (James, 1890, p. 244). The modern equivalent might be trying to open the fridge door quickly enough to see whether the light is always on (Blackmore, 2012; O'Regan, 2011). Similarly, asking 'Am I conscious now?' or 'What am I conscious of now?' or even 'What is phenomenality?' or 'What is consciousness?' can feel like turning on a light, but is that light always on? And if not then what is the darkness like?

With a fridge we can find out, for example by drilling a hole in the side and looking in when the door is still closed, or by understanding how the switch works. To some extent we can do this with brains too. We can look inside with electrodes or scanners and see what is going on.

We do not find a light and a switch or a cold dark cupboard full of food. But nor do we find a conscious self, a place where a conscious self could be, or any roles for it to play. We find billions of neurons connected up in trillions of ways with vast numbers of parallel processes going on simultaneously. There is no central processor, no place for an inner observer commanding the action, and no show in the Cartesian theatre (Dennett, 1991). To all appearances there are just lots and lots of neurons firing and chemicals moving about. If we ask which are the conscious ones, how can we tell?

Can we take a different tack and look into the darkness personally, by training our introspection to look more carefully? I have described in detail my many attempts to do this using meditation combined either with formal Zen koan practice or with persistently asking questions of my own devising (Blackmore, 2011). Hundreds of students on my consciousness courses have explored these questions too (Blackmore, 2010/11). Once they get used to the sensation of becoming more conscious, or even 'waking up', by asking, 'Am I conscious now?', a second question naturally pops up: 'Was I conscious a moment ago?' or 'What was I conscious of a moment ago?'. Asking these, I suggest, leads to a curious discovery — that we ourselves do not know the answer.

A striking example of this occurred just yesterday. I was climbing a steep hill with rough and unequal steps, noticing the changing rhythm as I climbed; one left, two right, two left, one right. I asked myself (as I so often do), 'Am I conscious now?' (yes, I'm conscious of the climbing) and then, 'What was I conscious of a moment ago?'.

Obviously there was the rhythm of climbing, but once I had asked the second question I could also remember hearing — or having had the feeling that someone or something had been hearing — the scrunching of my feet on the rough surface which I had not noticed until I asked. Then I remembered hearing my own laboured breathing going in and out, the effort in my legs, and the burning sensation of the sun on my shoulders. In an odd way these now seemed to have been as conscious as the rhythm, although in another way they seemed not to have been conscious at all because I had only just noticed them. I was musing briefly on this familiar and long-practised oddity when suddenly, in a startled flash, I remembered something else.

Just two steps before, I had lifted my arm, looked at my watch, seen that it was just before noon, decided that I was on track for where I was going, and dropped my arm again. The memory was clear, detailed, and vivid. So was I conscious of checking the time?

I might answer yes, because now I could clearly remember doing it, even to the look of my arm in the sun and the position of the hands on my watch. I might answer no, because the whole memory seemed to come 'into my consciousness' afterwards only because I asked the second question. Which is right? I do not know, and if I do not know then how can the answer be discovered? Indeed, is there an answer at all?

I suggest not. To explore further, let me ask some other, related, questions. Did I consciously decide to check the time? I had no memory of intending or planning the action but I might just have forgotten the intention. How can I find out?

Was performing the actions conscious at the time? The same applies.

Here's a harder, but important, one. Were the brain processes underlying the actions conscious brain processes? Clearly the actions are quite complex and involved multiple perceptual, cognitive, and motor processes, just as noticing the rhythm of climbing the steps does. Can we find out which of these were conscious or unconscious processes?

We could try.

We might look to the concept of attention. If attention is thought of as resource allocation then considerable attention must have been paid to looking at my watch and registering the time, and presumably it would be possible, at least in principle, to measure the amount of work going into all this. So, although I seemed to be conscious only of the rhythm of climbing, resources were clearly split, as they are in the familiar 'unconscious driving phenomenon', and it would be hard to say that far more resources were being used for one than the other. This means that we cannot use the amount of attention in any simple way to decide which processes were conscious and which were unconscious.

We might look to the popular global workspace theories for an answer. According to Baars' (1988) original formulation of GWT the contents of consciousness are the processes currently in the GW, or 'on the stage', and they become conscious by virtue of being globally broadcast to the rest of the unconscious brain. According to the later 'neuronal GWT', information becomes conscious when the long-distance connectivity of 'workplace neurons' makes it widely available and this is what we experience as a conscious state (Dehaene and Naccache, 2001). But what does this mean? And how and why does this broadcast make information conscious?

There are two radically different ways of interpreting GWTs. The first, and more common, is that when the contents on the stage are broadcast they 'become conscious'. This version of Cartesian materialism (Dennett, 1991) retains the hard problem. Something magical has happened to the previously unconscious contents so that they are now conscious ones. They have gained, or have become, qualia or phenomenality. The alternative is that nothing more happens to them at all. Being broadcast is all there is. This is what Dennett (2001) means by 'cerebral celebrity' or 'fame in the brain'. Just as there is nothing more to being famous than being well known by lots of people, so there is nothing more to consciousness than being widely available in the brain. This availability has consequences for later actions and perceptions, including the ability to talk about what happened, to attribute consciousness to the sensations, and to base further actions upon them.

Applying GWT to my example, we might say that before I asked the question the only processes on the stage or in the GW were those involved in climbing and counting the steps; these were being broadcast to the rest of the unconscious brain while those involved in checking the time were not.

Now we can see the problem. Had I not asked the question I would have totally forgotten the way my arm looked as it rose and fell. Yet this action would certainly have had consequences for later brain processes, thoughts, and actions. If, when I arrived at the top of the hill, I had wondered what time it was I would have known that it was just gone twelve. So what are we to say? That the broadcast from looking at my watch was, for some reason, not sufficient or not of the right sort of broadcast to count as being 'on the stage' or 'in consciousness'? If so, for what reason?

I suggest that the whole idea of the GW, and its popularity, arises from the illusion I have described. Whenever we ask about consciousness, a temporary unity of a set of thoughts and perceptions is constructed and is linked to a representation of self as a continuing observer (Metzinger, 2009). This we call the contents of our consciousness while everything else is called 'unconscious'. GWT nicely captures this intuition, which is based on a momentary and misleading situation.

Returning to my example, as I asked the first question a self-model was briefly constructed of me climbing, counting, and looking at the ground, but not including looking at my watch. If we could look inside the brain in sufficient detail I guess we would see some of the hill-climbing processes linked to the body schema, and to self-modelling and questioning processes while the watch-looking and many other processes were going on separately. When I asked the second question, lots more processes, including the watch-looking, were combined to make an even more complex whole. When I stopped asking about consciousness and got on with climbing the hill the temporary coherence dissolved and normality resumed. The multiple parallel processes just carried on, none linked to a model of self as observer; none either in or out of consciousness; none either conscious or unconscious.

I conclude that there is no intrinsic difference between conscious and unconscious processes, nor between conscious and unconscious actions or perceptions. Rather, consciousness is a fleeting attribution that we make if and when we ask about it, either when asking such questions as 'What am I conscious of now?' or in retrospect when we think about the past. This implies that most theories of consciousness address only rare moments in our lives.

So what was it like before we asked? We might try to find out using either the methods of neuroscience or of disciplined introspection. Amazingly enough, both come to the same answer: that there are lots and lots of parallel processes going on and no obvious way to tell which were conscious.

Is it possible to find out? Are there actual, but unobservable, facts about which processes, thoughts, or actions were conscious at any time? I say no. The neural correlates of all these would in principle be observable but with no way of distinguishing between conscious and unconscious ones either by objective measures or in subjective experience. The distinction is meaningless because consciousness is an attribution we make, not a property of events, thoughts, brain processes, or anything else.

If this is so, we must reject not only many folk-psychological beliefs but also many common phrases used in the literature, with interesting implications for the science of consciousness.

2. Implications for the study of consciousness

2.1. There is no stream of consciousness

William James coined the phrase 'the stream of consciousness' in his classic 1890 work *The Principles of Psychology* and it has been used ever since to imply something like the folk-psychological idea that we are conscious subjects experiencing an ever-changing but unified flow of ideas, thoughts, perceptions, and intentions. I suggest that consciousness appears as a stream only when we reflect on it as such. The rest of the time, multiple parallel processes carry on, sometimes interacting with each other, often not. When we ask 'What am I conscious of now?' or 'What is it like being me now?' some are gathered together and the answer appears stream-like. There are memories of recent perceptions and thoughts and, if we remain mindful for a few moments, a changing array of new perceptions and thoughts coming along. There is a powerful sense of someone who is experiencing this stream. The illusion is to believe there is also a stream like this when we are not enquiring (Blackmore, 2002).

2.2. There are no contents of consciousness

The idea that consciousness has contents is perfectly natural and easy to imagine. It appears in many guises; as the contents of the stream, the items on the stage of the GW, the features illuminated by the spotlight of attention, and the show in Dennett's mythical Cartesian theatre.

There is a new problem revealed here. I have used the words 'item', 'feature', and 'show' but what is really being referred to? When talking in brain terms people often say 'processes', 'representations', 'neural patterns', or just 'information': in psychological terms, 'items', 'ideas', 'perceptions', 'thoughts', or even just 'things'. I find myself reverting to 'things' when struggling to understand what is supposed to be in the stream, in the GW, or on the stage.

Ignoring this problem, these 'things' must start by being unconscious, perhaps building up ready to enter the GW or moving into the spotlight of attention. Then they become conscious before finally leaving and becoming unconscious again, perhaps disintegrating as they do so. What could it mean for these things (items, processes, thoughts, information, or whatever) to become conscious? Do they change from being objective things to subjective experiences? Do they suddenly become phenomenal or qualia-laden (Ramachandran and

Hirstein, 1997) or get qualia attached (Gray, 2004)? The hard problem has not disappeared but has merely been restricted to 'contents' inside the hypothetical stream, GW, or stage. The impression of a space that contains everything we are conscious of at any time seems natural and obvious, but consciousness is not a container (Blackmore, 2002).

2.3. The unity of consciousness

This phrase is often taken to mean that at any time all the things I am conscious of form some kind of collective. It can certainly seem that way, as if the sensations of sitting here, the pictures on the wall in front of me, the sounds around me, all form some kind of unified field. We might guess that, in neural terms (this is, of course, an empirical question), attention brings together some of the ongoing parallel processes and so creates temporary coalitions that provide a sense of unity. Yet it takes only a little careful observation to see that as my attention shifts these sensations come and go and that these acts of attention pull some together and let others disappear. This is the only sense in which there is 'unity of consciousness'.

2.4. The continuity of consciousness

This is the easiest idea to demolish by trying to 'look into the darkness'. Every time I 'turn on the light' by asking myself what my consciousness is like, it seems to be a flowing stream of ever-changing, unified contents, much as it was last time I looked. That's fine. That's how it seems, and how it seems is what we are trying to explain.

The illusion is to leap from that repeated observation to the conclusion that consciousness is always that way; that the stream of contents continues without break when it does not.

2.5. The neural correlates of consciousness

The hunt for the NCCs is probably the most popular research paradigm in consciousness studies (Metzinger, 2000). From experiments in binocular rivalry to Baars' (1988) 'contrastive analysis', the idea is to find the difference between those neural processes that are conscious and those that are not. If my analysis is correct it is obvious that this whole approach is doomed.

I do not doubt that neuroscientists can find, in ever greater detail, the neural correlates of specific actions, thoughts, perceptions, and so on. They can also look for the neural correlates of asking such questions as 'What is it like to be me now?' or 'What is consciousness?' and of making attributions of consciousness to past or present thoughts and perceptions. But they will never find the neural correlates of an extra added ingredient — 'consciousness itself' — for there is no such thing.

2.6. The function of consciousness

We humans are conscious, so the argument goes, therefore consciousness must have evolved for a reason and must have a function. Many functions have been proposed, including error monitoring (Crook, 1980), an inner eye (Humphrey, 1986), saving us from danger (Baars, 1997), late error detection (Gray, 2004), and giving a point to survival (Velmans, 2009). Yet this type of argument can be like believing in the possibility (not the conceivability) of philosophical zombies. It assumes some special extra thing called consciousness 'itself' in addition to all the underlying functions and processes. It assumes that there might have been creatures with and without this special extra and that the ones who had it survived, reproduced, and passed on their genes more effectively than those without.

If my analysis is right then this makes no sense. Consciousness, as subjectivity, is not a force that can do anything or have effects or consequences that could form the basis of any evolutionary function. It did not evolve. What evolved was an intelligent creature with the capacity for selective attention, language, and introspection, the ability to call some of its own actions, thoughts, or perceptions conscious, and hence to fall for the illusion that consciousness has a function.

As Frankish points out, this offers a new perspective on the function of consciousness, because we can ask what function such illusions serve or how they might be adaptive. In his later work, Humphrey (2006; 2011) spells out one answer; that from monitoring their own responses to the world creatures internalized sensations, leading to an inner 'magical-mystery show' and an illusory soul. Consciousness seems so important to us because '*it is its function to matter*' and to seem other-worldly and mysterious (Humphrey, 2006, p. 131).

This is a radically different approach. Yet, like these other theories, it still assumes that the ultimate function is biological survival; that consciousness, or the illusions of consciousness, must benefit human genes. But this is not the only possibility.

Elsewhere I have argued that the benefit accrues more to memes than genes (Blackmore, 1999). To survive and reproduce memes need first to find homes in human brains and then find ways to get passed on. A meme that becomes 'my' idea, preference, need, favourite joke, or special song has an evolutionary advantage over one that does not. Memes that make up the stories I tell about myself or the opinions I express have an advantage, and they build up into what I have called the 'selfplex', a co-adapted complex of memes that all thrive better together than apart. A self with strong opinions, lots of ideas, and a need for status and power makes an effective meme spreader even if it is not the continuous, unified, and powerful subject of experience it models itself as being.

Is the illusory self, as Dennett (1991) would have it, a 'benign user illusion'? I think not. It is arguably a 'malign user illusion' and the source of much suffering and misery (Blackmore, 2010/11). This is the self who craves love, friendship, status, possessions, and power. This is the self who gets disappointed, hurt, lonely, angry, and resentful. This is the self who wants happiness but when happy fears losing it. This is the self who makes constant comparisons with others and fears other people's judgments. It is strange that an illusion can entail so much suffering.

Is it possible to throw off these illusions completely? Mystics and contemplatives for thousands of years have claimed that it is; that greed, hatred, and delusion can give way to kindness and equanimity, even if the endeavour takes decades of meditation or years of solitude in a remote mountain cave. Others have discovered similar insights through spontaneous mystical experiences or the use of shamanic brews and psychedelic drugs. Many long-term meditators describe experiences without an experiencer, or non-dual states in which experience and experiencer are one. In these states we can ask 'Am I conscious now?' and truthfully answer 'no' because there is no constructed self to be the experiencer (Blackmore, 2011). The experienced duality of self and other then disappears along with the hard problem and yet it is hard to make rational sense of this directly experienced non-duality.

3. Conclusion

As Frankish suggests, the question is not whether illusionism is intuitively plausible, but whether it is rationally compelling. His illusionism aims to explain why phenomenal consciousness *seems* to exist. My own aim is the slightly different one of using introspection to explain why some other delusions about consciousness are so compelling, and perhaps this way make delusionism more intuitively plausible. Mysteries remain and, as Dennett puts it, a mystery is something we don't yet know how to think about. All I have tried to do is to clear away some of the intuitively appealing but wrong ways of thinking about consciousness in the hope that they may be replaced by better ones.

References

Baars, B.J. (1988) *A Cognitive Theory of Consciousness*, Cambridge: Cambridge University Press.

Baars, B.J. (1997) *In the Theatre of Consciousness: The Workspace of the Mind*, New York: Oxford University Press.

Blackmore, S. (1999) *The Meme Machine*, Oxford: Oxford University Press.

Blackmore, S. (2002) There is no stream of consciousness, *Journal of Consciousness Studies*, **9** (5), pp. 17–28.

Blackmore, S. (2010/11) *Consciousness: An Introduction*, 2nd ed., London: Hodder Education, and 2011 New York: Oxford University Press.

Blackmore, S. (2011) *Zen and the Art of Consciousness*, Oxford: Oneworld Publications.

Blackmore, S. (2012) Turning on the light to see how the darkness looks, in Kreitler, S. & Maimon, O. (eds.) *Consciousness: Its Nature and Functions*, pp. 1–22, New York: Nova.

Blackmore, S., Brelstaff, G., Nelson, K. & Troscianko, T. (1995) Is the richness of our visual world an illusion? Transsaccadic memory for complex scenes, *Perception*, **24**, pp. 1075–1081c.

Crook, J. (1980) *The Evolution of Human Consciousness*, Oxford: Oxford University Press.

Dehaene, S. & Naccache, L. (2001) Towards a cognitive neuroscience of consciousness: Basic evidence and a workspace framework, *Cognition*, **79**, pp. 1–37.

Dennett, D.C. (1991) *Consciousness Explained*, London: Little, Brown & Co.

Dennett, D. (2001) Are we explaining consciousness yet?, *Cognition*, **79** (1), pp. 221–237.

Gray, J. (2004) *Consciousness: Creeping Up on the Hard Problem*, Oxford: Oxford University Press.

Humphrey, N. (1986) *The Inner Eye*, London: Faber & Faber.

Humphrey, N. (2006) *Seeing Red*, Cambridge, MA: Harvard University Press.

Humphrey, N. (2011) *Soul Dust: The Magic of Consciousness*, Princeton, NJ: Princeton University Press.

James, W. (1890) *The Principles of Psychology*, London: Macmillan.

Levin, D.T., Momen, N., Drivdahl IV, S.B. & Simons, D.J. (2000) Change blindness blindness: The metacognitive error of overestimating change-detection ability, *Visual Cognition*, **7** (1–3), pp. 397–412.

Metzinger, T. (ed.) (2000) *Neural Correlates of Consciousness*, Cambridge, MA: MIT Press.

Metzinger, T. (2009) *The Ego Tunnel: The Science of the Mind and the Myth of the Self*, London: Basic Books.

O'Regan, J.K. (2011) *Why Red Doesn't Sound Like a Bell*, New York: Oxford University Press.

O'Regan, J.K. & Noë, A. (2001) A sensorimotor account of vision and visual consciousness, *Behavioral and Brain Sciences*, **24** (5), pp. 939–1011.

Ramachandran, V.S. & Hirstein, W. (1997) Three laws of qualia: What neurology tells us about the biological functions of consciousness, *Journal of Consciousness Studies*, **4** (5–6), pp. 429–457.

Simons, D.J. & Rensink, R.A. (2005) Change blindness: Past, present, and future, *Trends in Cognitive Sciences*, **9** (1), pp. 16–20.

Velmans, M. (2009) *Understanding Consciousness*, London: Routledge.

Daniel C. Dennett

Illusionism as the Obvious Default Theory of Consciousness

Abstract: *Using a parallel with stage magic, it is argued that far from being seen as an extreme alternative, illusionism as articulated by Frankish should be considered the front runner, a conservative theory to be developed in detail, and abandoned only if it demonstrably fails to account for phenomena, not prematurely dismissed as 'counter-intuitive'. We should explore the mundane possibilities thoroughly before investing in any magical hypotheses.*

Keith Frankish's superb survey of the varieties of illusionism provokes in me some reflections on what philosophy can offer to (cognitive) science, and why it so often manages instead to ensnarl itself in an internecine battle over details in ill-motivated 'theories' that, even if true, would be trivial and provide no substantial enlightenment on any topic, and no help at all to the baffled scientist. No wonder so many scientists blithely ignore the philosophers' tussles — while marching overconfidently into one abyss or another.

The key for me lies in the everyday, non-philosophical meaning of the word *illusionist*. An illusionist is an expert in sleight-of-hand and the other devious methods of stage magic. We philosophical illusionists are also illusionists in the everyday sense — or should be. That is, our *burden* is to figure out and explain *how the 'magic' is done*. As Frankish says:

> Illusionism replaces the hard problem with the illusion problem — the problem of explaining how the illusion of phenomenality arises and

Correspondence:
Daniel C. Dennett, Tufts University, Medford, MA, USA.
Email: daniel.dennett@tufts.edu

why it is so powerful. This problem is not easy but not impossibly hard either. The method is to form hypotheses about the underlying cognitive mechanisms and their bases in neurophysiology and neuroanatomy, drawing on evidence from across the cognitive sciences. (p. 37)

In other words, you can't be a satisfied, successful illusionist until you have provided the details of how the brain manages to create the illusion of phenomenality, and that is a daunting task largely in the future. As philosophers, our one contribution at this point can only be schematic: to help the scientists avoid asking the wrong questions, and sketching the *possible* alternatives, given what we now know, and motivating them — as best we can. That is just what Frankish has done.

He distinguishes and assesses the different versions of illusionism, illuminating the paths that led to them. His account of my own view is flawless, and I was enlightened by many of his remarks about the ideas of other illusionists I thought I fully understood. He is particularly good on the various motivations of the variations, but I want to stress the motivation for the overall strategy of taking illusionism — in any form — seriously. And here, once again, it is useful to take a glance at stage magic. In today's world, if not in the Dark Ages or even the Renaissance, the standard, default assumption about any feat of stage magic we encounter is that it is (somehow) the product of everyday physical causes, involving perhaps magnets and electricity, or even holograms, but not psychokinesis, clairvoyance, or the assistance of any poltergeists, goblins, or other supernatural agents. In other words, it is *stage* magic, not 'real magic'. (In his excellent book on Indian street magic, *Net of Magic: Wonders and Deceptions in India*, Lee Siegel writes,

> 'I'm writing a book on magic,' I explain, and I'm asked, 'Real magic?' By *real magic* people mean miracles, thaumaturgical acts, and supernatural powers. 'No,' I answer: 'Conjuring tricks, not real magic.' *Real magic*, in other words, refers to the magic that is not real, while the magic that is real, that can actually be done, is *not real magic*. (Siegel, 1991, p. 425))

The common understanding is that magicians are doing tricks, not engaging in sorcery, and moreover, the default — but defeasible — assumption about any feat of stage magic is that it doesn't avail itself of quantum entanglement (Einstein's 'spooky action at a distance') or the creation of any new subatomic particles or other forces that would institute a revolution in physics. Hey, it's just stage magic!

In short, when it comes to stage magic we assume, until positively shown otherwise, that the effects are achieved by some hard-to-imagine concoction of everyday physical causes and effects. Here is where anybody, philosopher or scientist or visionary, is apt to suffer a failure of imagination and mistake it for an insight into necessity. As the noted illusionist Jamy Ian Swiss has said, 'No one would ever think that we would ever work this hard to fool you. That's a secret, and a method of magic' (2007, the e.g.conference, www.egconf.com/videos/how-magic-works). This is not just an interesting observation. It draws attention to a fact that puts *all* philosophers on notice: nobody would, or should, take seriously a would-be explainer of stage magic who declared that it was just undeniably, intuitively obvious that no *possible sequence of ordinary physical events* could account for the feat just observed. We philosophical illusionists say that before you run off half cocked with theories about consciousness as one sort or another of 'real magic', you should try to explain it all as an illusion engendered by nature. When philosophers rely on what follows from their most undeniable *intuitions* about consciousness, before exhausting the physical possibilities, they are not adopting a sound method of enquiry but simply indulging themselves. Imagine Chalmers' declaration that phenomenal consciousness is a datum, transposed into the claim that a lady-sawn-in-half is a datum, or the claim that we are directly acquainted with the real presence of a lady-sawn-in-half. You may *think* you're directly acquainted with this, but that's a fact of personal psychology, at best an unshakable intuition, not a datum. You could be wrong, and until we have canvassed the alternatives, we should put your intuition on the back burner, not honour it. Or consider Searle's italicized dictum that '*where consciousness is concerned the existence of the appearance is the reality*' (quoted by Frankish, this issue, p. 32). Maybe, and maybe not. Searle apparently thinks that this is crushingly obvious, and he is not alone. When we know more about the brain's activities we *will see* if we can eliminate the prospect of the brain creating an illusion of 'appearance', of phenomenality. You can't just declare, as a first principle, that this is impossible.

Illusionism may today seem 'incredible, even ludicrous' (Frankish, p. 37), but if and when it is eventually fleshed out with details, those gullible folks who fell for the trick will discover that they have industriously concocted their card castle 'theories' for nothing. I put 'theories' in scare quotes because most philosophical theories are just definitions defended, with no aspiration to make novel predictions but rather just to assign the phenomena covered by the 'theory' to some category or other. They at best clarify and articulate the implications of the everyday concepts involved. A weakness of such 'theories' is that, since they are largely driven by shared folk intuitions, they are always playing catch-up, seeing if they can accommodate newly discovered but unanticipated scientific discoveries, instead of pioneering perspectives from which new empirical questions can be asked and answered. When counter-intuitivity counts against an hypothesis, with nothing but consistency to other intuitions to support it, there is little chance of making progress. Illusionism boldly discards a host of all too comfortable intuitions. As Frankish says, 'The question is not whether illusionism is intuitively plausible, but whether it is rationally compelling' (p. 37).

Illusionism, I am saying, should not be seen as a lame attempt to deny the obvious, but as the leading contender, the default view that should be assumed true until proven otherwise. (I grant that my whimsical title, 'Quining Qualia', lent unintended support to the perception that illusionism is a desperate and incredible dodge, and for that little joke I now repent.)

From this perspective, we can see that philosophers of mind who are *not* illusionists are prematurely encouraging scientists to worry about the wrong questions, artefactual problems like those that would arise for any scientists trying to uncover the details of the quantum-entanglement theory of teleporting the beautiful assistant from one trunk to another or trying to reconcile the actual presence of the ten of diamonds in their pocket where they put it with its manifest presence on the table. It is a remote possibility that we will have to fall back on quantum physics or multiple universes to account for some mind-boggling bit of magic, but first, let's try the conservative route.

One of the difficulties illusionists must tackle is how to temporize with terminology until the facts are in, and here the big weak spot, in my opinion, is the term 'representation'. A fanatic about proper procedure would insist that one *never* use the term without saying, very clearly and in detail, who or what the user/audience is for the so-called representation. But we simply cannot meet that burden *in detail* yet. So we are tempted to leave the 'end user' slot unspecified, which amounts, in theories of consciousness, to evading what I call the Hard Question: 'And then what happens?' (Dennett, 1991, p. 255). It makes a big difference whether a postulated representation is *for* modulating hand–eye coordination or *for* identification or categorization of objects, or *for* ... conscious experience. Today we — most of us —

are comfortable with systems of *unconscious* representations that influence, specify content, orient, direct memory retrieval, etc. That is as good as gospel in cognitive science. These are representations *in us* that contribute to our cognitive talents without being *for us.* (In this regard they are no different from the representations of blood sugar level or vitamin deficiency that modulate our digestive systems without engaging cerebral cortex at all.) But at some point, as Frankish puts it, we must describe

> the sensory states that are the basis for the illusion. On most accounts, I will assume, these will be representational states, probably modality-specific analogue representations encoding features of the stimulus, such as position in an abstract quality space, egocentric location, and intensity. (p. 19)

Filling in these details will require answering a host of questions that Frankish raises:

> Is introspection sensitive only to the content of sensory states, *or* are we also aware of properties of their neural vehicles? Do the reactions and associations evoked by our sensory states *also* contribute to the illusion of phenomenality?... Are sensory states continually monitored *or* merely available to monitoring? Is the introspectability of sensory states a matter of internal access and influence *rather* than internal monitoring? (p. 19, my italics)

I submit that, when we take on the task of answering the Hard Question, specifying the uses to which the so-called representations are put, and explaining how these are implemented neurally, some of the clear alternatives imagined or presupposed by these questions will subtly merge into continua of sorts; it will prove not to be the case that content (however defined) is sharply distinguishable from other properties, in particular from the properties that modulate the 'reactions and associations evoked'. The distinction between continual monitoring and availability to monitoring will also vanish (it already has vanished in many computer programs). Suppose that neural signals with content (*consistent with*) *dog here now* arrive at location X in the cortex every 20 milliseconds, thereby preventing location X from *enquiring* whether there seems to be a dog here now. Is location X continually monitoring for dog-presence, or just lulled into a complacent state of disinterest? (*Cf.* in what sense is 'the background' *there* in unattended 'phenomenal consciousness'?) This may not be a well-motivated question when we learn more about the mechanisms involved (see Cohen, Dennett and Kanwisher, 2016). Frankish's questions are good questions, but that doesn't mean that they will all have crisp answers. The answer may well be that these distinctions do not travel well when we abandon what Sellars (1962) calls the manifest image and get down in the trenches of the scientific image.

Here is a sentence that tantalizes me, in Daniel Wegner's book, *The Illusion of Conscious Will*:

> We can't possibly know (let alone keep track of) the tremendous number of mechanical influences on our behavior because we inhabit an extraordinarily complicated machine. (Wegner, 2002, p. 27)

This frankly Cartesian formulation exposes the allure of the traditional manifest image: there is a place in the brain where 'our minds portray their operations to us, then, not their actual operation' (*ibid.*, p. 96). Wegner was silent about how this portrayal (to 'us') is accomplished, so it is not clear if he counts as an illusionist. I think he should count, since he gave a good if partial answer to the Hard Question while eschewing all questions about the 'phenomenality' of the portrayals. As he recognized, the effects he provoked in his experiments could be accounted for as stage magic.

The seeds of illusionism can already be discerned in U.T. Place's pioneering article, 'Is Consciousness a Brain Process?' (1956). Place was so bold as to identify the *denial* of illusionism as a fallacy, the *phenomenological fallacy*:

> [T]he mistake of supposing that when the subject describes his experience, when he describes how things look, sound, smell, taste, or feel to him, he is describing the literal properties of objects and events on a peculiar sort of internal cinema or television screen, usually referred to in the modern psychological literature as the 'phenomenal field'. (*ibid.*)

How then does one avoid the phenomenological fallacy? J.J.C. Smart offered an answer in 1959 and elaborated on his answer in 1963:

> The man who reports a yellowish-orange after-image does so in effect as follows: '*What is going on in me is like what is going on in me when my eyes are open, the lighting is normal*, etc., etc., and there really is a yellowish-orange patch on the wall.' (Smart, 1963, p. 94)

As Smart pointed out way back then, it is quite possible for a mechanism to be able to discern or discriminate when something going on in it is like something else without having any idea just wherein that similarity resides. If we fill the heads of people with such mechanisms, suitably organized and orchestrated, they can provide a large part of the answer to 'And then what happens?' without ever postulating anything like phenomenality.

How might this go? When you seem to see a red horizontal stripe (as a complementary-colour after-image of a black, green, and yellow American flag), there is no red stripe in the world, no red stripe on your retina or in your brain. There is no red stripe anywhere. There is a 'representation' of a red stripe in your cortex and this cortical state is the source, the cause, of your heartfelt conviction that you are in the presence of a red stripe. You have no privileged access to how this causation works. We have a good theory of how colour perception works, with its opponent processes and refractory periods, so you can probably explain the early or distal links in the causal chain from eyeball to conviction, but you simply don't know what the proximal or immediate causes are that put you into a state of subjective conviction and the attendant further sequelae ('and then what happens?'). (And this is true of your access to normal, not illusory, vision as well, of course.)

The red stripe you seem to see is not the *cause* or *source* of your convictions but the *intentional object* of your convictions. In normal perception and belief, the intentional objects of our beliefs are none other than the distal causes of them. I believe I am holding a blue coffee mug, and am caused to believe in the existence of that mug by the mug itself. The whole point of perception and belief fixation is to accomplish this tight coalescence of causes and intentional objects. But sometimes things go awry. Suppose a gang of hoaxers manage to convince you, by a series of close encounters, that there is a space alien named Zom who visits you briefly, speaks to you on the phone, etc. The *causes* of your various Zom experiences can be as varied as can be, so that nothing at all in the world deserves to be identified as Zom, the intentional object of your beliefs.

What are intentional objects 'made of'? They're not made of anything. When their causes don't coalesce with them, they are fictions of a sort, or illusions. We don't postulate *fictoplasm* as the substance from which Sherlock Holmes and Hamlet are composed, and likewise we shouldn't postulate *figment* as the stuff of 'phenomenal properties' (Dennett, 1991). When there is a red stripe in the world, the redness is a complex physical property of the stripe; when there just seems to be a red stripe in the world, that very same property is *represented* as being present by some team of brain agents that are the cause of your false conviction. The eternally tempting mistake is to fall for this chain of inference:

It seems to me as if there's a red stripe projected out onto that wall, but there is no such red stripe out there,... *so it must be in here.*

And even when there is a red stripe out there in the world that I see, my seeing it must involve an intermediary 'phenomenal' red stripe in my consciousness.

We illusionists advise would-be consciousness theorists not to be so confident that they *couldn't* be caused to have the beliefs they find arising in them by mere neural representations lacking all 'phenomenal' properties. Of course they could; just ask stage magicians — illusionists in the everyday sense — who specialize in provoking false but passionately held beliefs in things that they seemed to see but didn't see.

References

Cohen, M., Dennett, D.C. & Kanwisher, N. (2016) What is the bandwidth of perception?, *Trends in Cognitive Sciences*, **20** (5), pp. 324–334.

Dennett, D.C. (1991) *Consciousness Explained*, Boston, MA: Little, Brown.

Place, U.T. (1956) Is consciousness a brain process?, *British Journal of Psychology*, **47**, pp. 44–50.

Sellars, W. (1962) Philosophy and the scientific image of man, in Sellars, W., *Science, Perception and Reality*, London: Routledge & Kegan Paul.

Siegel, L. (1991) *Net of Magic: Wonders and Deceptions in India*, Chicago, IL: University of Chicago Press.

Smart, J.J.C. (1959) Sensations and brain processes, *The Philosophical Review*, **68** (2), pp. 141–156.

Smart, J.J.C. (1963) *Philosophy and Scientific Realism*, London: Routledge & Kegan Paul.

Swiss, J.I. (2007) A lecture at the e.g.conference, [Online], www.egconf.com/videos/how-magic-works.

Wegner, D.M. (2002) *The Illusion of Conscious Will*, Cambridge, MA: MIT Press.

Jay L. Garfield
Illusionism and Givenness

Abstract: *There is no phenomenal consciousness; there is nothing 'that it is like' to be me. To believe in phenomenal consciousness or 'what-it's-like-ness' or 'for-me-ness' is to succumb to a pernicious form of the Myth of the Given. I argue that there are no good arguments for the existence of such a kind of consciousness and draw on arguments from Buddhist philosophy of mind to show that the sense that there is such a kind of consciousness is an instance of cognitive illusion.*

Frankish argues that phenomenal consciousness is an illusion. I think he is right. Here I adduce a few other considerations in support of the thesis that — in the sense relevant to those who believe that there are phenomenal properties or concepts, or who believe that there is 'something that it is like' to be conscious, or that there is immediate experience, or that there are qualia, or that there is a 'for-me-ness' of experience; in short for anyone who subscribes to any of the currently popular forms of the Myth of the Given — there is no phenomenal consciousness.

   I will begin by querying the sense of the popular phrase introduced by Nagel (1974), 'what it is like', arguing that there is nothing that it is like to experience something, nothing that it is like to have qualitative experience. I will then ask whether the distinction between sentient beings and zombies demanded by those who believe in phenomenal properties makes any sense, arguing that if zombies are possible, we are all zombies. I will then argue that we can make no sense of any knowledge of qualitative properties. I will conclude by arguing that neither introspection nor transcendental argument can give us any reason to believe that there is anything answering to the demands of

Correspondence:
Jay L. Garfield, Smith College, Northampton, MA, USA; Harvard Divinity School; University of Melbourne; Central University of Tibetan Studies.
Email: jgarfield@smith.edu

qualitative consciousness. The very idea that there is an inner world of qualitative states must be illusory.

1. That there is nothing that it is like to be me

Nagel has a lot to answer for. Generations of philosophers have grown up believing not only that there must be something that it is like to be a bat, but that there is something that it is like for each of us to be *us*, and that that *something* is our qualitative consciousness. It is but a short step to analysing that consciousness is the immediate apprehension of qualia or qualitative properties, and of course only a short step from there to a view of perceptual experience on which we experience the external world only in a mediate fashion, mediated by the immediate experience of an inner world. And from there, one more short step takes us to idealism. So, it is best to avoid the *proton pseudos*.

There is something that a yellow mango is like. It is yellow, oblate, sweet, etc... To say that it is like that is to ascribe to it perceptible properties. There is something that middle C played on an oboe is like. It is a tone of a particular timbre and frequency. Again, to say what something is like is to list its perceptible properties. We focus on the object, and we characterize it. In short, to the extent that that experience has a subject–object structure, when we say what something is like, we characterize the objective, not the subjective, pole of experience.[1] (A bit later in this essay, I will consider the more radical possibility that the very thematization of experience in terms of subject–object duality is itself a cognitive illusion.)

Bats — as Nagel correctly observes — use sonar to perceive the world. Their objects are thus perceived via a sensory modality that we lack, and so they perceive sensible properties that we do not. If bats could speak, or characterize the world they experience, they could say what e.g. a moth is like. And they would characterize it differently from the way we would, because they are sensitive to different perceptible properties. But what they would tell us is what *a moth is like*, not *what it is like to perceive a moth via sonar,* just as when I describe a moth to a talking bat, I would say what a moth *looks like,* not *what it is like to see a moth with eyes.*

---

[1] This is not to say that we are in no way aware of our interior lives. We use proprioception, nociception, and interoception to become aware of states of our bodies. But these are systems that give us access to states of our bodies, not to *qualia*.

Nagel's sleight of hand — the decisive move in the conjuring trick, as Wittgenstein put it (1953, §308) — is to convert *what the world is like for a bat* to *what it is like to be a bat*. That is to confuse the object with the subject, and to ask us to assign properties to our subjectivity, as opposed to the objects we experience. It is easy to accept that invitation, since it is almost irresistible to think of our experience as constituting an inner domain populated by inner particulars that constitute the immediate objects of our experience, and contrasting with an outer domain of objects we know only indirectly. But it is not so easy to preserve cogency once we succumb to this temptation. For when we reach for predicates to characterize our experience, the only ones we have are those that characterize the objects of our experience, not the subjects. When I have described the mango to you, I have said all I can ever say about my experience of it.

If you ask me what it is like to be me, *simpliciter*, you ask me to describe consciousness itself, apart from any object of consciousness. There is no such thing. Consciousness is always consciousness *of* something, and when the object is subtracted, nothing remains to be characterized. Now, some might say that in talking of experience in this way I am reducing our consciousness to that of zombies — beings just like us, but who lack a phenomenal awareness. I agree, but, I will argue, this is not a *bad* thing.

2. I am a zombie

It is important to the picture of perceptual acquaintance with the world according to which it is mediated by phenomenal consciousness that to perceive the world *with* phenomenal consciousness is different from what it would be to perceive the world *without* phenomenal consciousness. Otherwise, to say that we are phenomenally conscious is just to say that we perceive, and the term would be idle. This is often put as the difference between zombie perception and sentient perception. Phenomenal realists often urge that sentient beings — those it is something for it to be like — differ from zombies — those whom it is nothing to be like — precisely because our perception is mediated by phenomenal consciousness and theirs is not. It is therefore essential to this picture that zombies are possible, or at least conceivable.

Zombies so described, however, as I argue in Garfield (1996 and 2016), are inconceivable. Or, to put it another way, if they are genuinely conceivable, we are zombies. Briefly, here is the point: zombies are, *ex hypothesi*, functionally identical to sentient human beings, with all of the same beliefs as sentient human beings, *including perceptual beliefs and beliefs about their supposed phenomenal consciousness*. But these latter beliefs are false: they have no phenomenal consciousness, despite believing that they do.

The crucial point here is functional identity. The alleged possibility of zombies only gets us the reality of phenomenal consciousness over and above mere perceptual experience if zombies are functionally identical to us; the mere possibility of beings that are psychologically different from us tells us nothing at all. We already know that there are dogs and chimpanzees, after all. But if we are functionally identical to zombies, our beliefs have the same typical causes and effects, including our beliefs about our perceptual states and phenomenal consciousness. Since we, *ex hypothesi*, believe that we have qualitative states and phenomenal consciousness, so do zombies. Since we are functionally identical, those states have the same typical causes; so, either zombies, like us, have phenomenal consciousness, in which case they are not zombies, or we, like them, falsely believe that we do, in which case we are. Either way, there is no argument from the possibility or conceivability of zombies to the reality of phenomenal consciousness.

To sum up so far: the claim that we are phenomenally conscious is meant to distinguish us from beings who lack this attribute, but who are like us in other respects. It cannot do so, and so, for beings who are doxastically and epistemically like us, the very idea of phenomenal consciousness is empty of real content. Moreover, the idea that there is some such property draws its plausibility from the idea that our perceptual experience involves direct contact with qualitative states or properties that enable us to know external properties. There is no good reason to believe in such things. We now turn to positive reasons to believe that nothing could play the role in our epistemic lives they are meant to play.

3. If there were qualitative properties, we could never know them

I will now argue directly for the epistemic idleness of the entire idea of phenomenal consciousness, qualia, qualitative character, the what-it's-like-ness of subjectivity, or any other member of this cluster of pseudo-concepts. The phenomenal realist takes these to be the objects and properties we know most immediately, with the greatest certainty, those whose presence in our inner lives we cannot possibly doubt.

While our ostensible perception of external objects might be illusory, due to hallucination, non-ideal perceptual situations, etc., it is argued, the *experience* of perception is immune from such illusion. If there is such experience, there must be a class of objects of that experience, *viz.* phenomenal states and properties, and a mode of consciousness of those states and properties, *viz.* phenomenal consciousness.

There are at least two problems with this idea. First, as we just saw, to have knowledge of this kind would be to know that we are not zombies, and there is no way that we could ever know this. Since there is nothing that distinguishes us from zombies *epistemically*, and since if these properties or inner objects of experience were real they would be knowable, they are not knowable, and hence are not real phenomena at all. When we experience the blueness of the sky, the blueness of the sky is the only object of knowledge, not any inner blue phenomena, not inner blue phenomenal properties, let alone what it is like to experience blue, only the blueness.

Second, if we were to have *knowledge* of phenomenal experience, there would have to be some kind of account of how that knowledge could come about, and we immediately run afoul of the normative dimensions of knowledge itself. Our knowledge of the external world is enabled by the fact that the external world is public, and that the criteria for the application of terms and for the correctness of claims are public. This enables us to be corrected, and to use terms and sentences in ways determined by norms.

The sphere of knowledge can therefore only exist in the space of reasons; the space of reasons is constituted by publicly enforced norms of enquiry, assertion, and language use. And those norms require that the relevant objects of knowledge themselves be public. But the experience, posited by the phenomenal realist, is private. It is hence not even a candidate for knowledge; nor are its putative objects candidate objects of knowledge. But once again, if it is unknowable, it is entirely idle. If we have phenomenal consciousness, we could never know it; whatever it is we confuse with phenomenal consciousness, if it is something of which we have knowledge, it is not phenomenal consciousness. (These points, of course, have been made with great force by Wittgenstein, 1953, and Sellars, 1963.)

4. The poverty of introspection

But, the phenomenal realist asks, don't we know our own experience and the properties that characterize it in introspection? Surely we can look inside and see what it is like to see blue, what it is like to be conscious? If we could not, we could never know anything external, since the external world is nothing for us if it is not given *for us* (Kriegel, 2007; 2009; Zahavi, 2005).

Once again, it is important to be clear about what phenomena we do know in perceptual experience. First and foremost, we know the objects around us and their perceptible properties — a mango and its yellowness and sweetness; middle C on the oboe and its timbre; sandalwood and its smell, etc. When we introspect to find the experiences that correspond to the awareness of the contents of perception, we can discover *that we are aware of a sweet, yellow mango, the sound of middle C, or the scent of sandalwood*, but nothing more is added; we discover not two objects of knowledge — the external and the internal medium that makes the external object known — but only the external object and the fact that we are aware of it. And to the extent that we are aware of the *fact* of our awareness, we do not have an additional inner phenomenal object, but only a higher-order attitude directed upon our current cognitive state (Garfield, 2015).

Whence the illusion that there is not only something that the world is like but also something that it is like to experience the way the world is? The source, at least at a first pass, is the failure to distinguish the subjective from the objective side of experience. The way this confusion occurs is nicely articulated by Sellars (1963) in 'Empiricism and the Philosophy of Mind'. We seek a common core of experience shared by taking an object to *be blue* and *to look blue*. We then take that common core to be a *common subjective state*, and we posit (in the old days) sense-data or sense impressions or (these days) qualia or qualitative properties of experience as constituting that common core. Those data, impressions, qualia, or properties do not fall on the object side of the subject–object divide, and so we place them neatly on the subject side. Since they are particulars or properties of particulars, there must be something that they are like, and so there must be something that subjectivity is like. Phenomenal consciousness is born.

As Sellars (1963) argues, however, there is a fallacy in this tale of origins. *Looking blue* is not epistemically prior to *being blue*. We do not first learn to identify *appearances* and then learn to extrapolate from them to identify the properties that are their typical distal causes. We first learn to identify blue things, and only later, when we become aware of the possibility of perceptual error, do we learn to talk of things *looking* blue. We learn to say that something *looks blue* as a way of registering our temptation to say that it *is blue*, but also of hesitancy regarding that attribution. To say that something *looks blue* is not a way of saying that we have an *experience of blueness*, but a way of attributing blueness to an external object, albeit with hesitation, or for reporting our *temptation* to attribute blueness, but our recusal from that attribution.

This is not, however, to deny that we have sensations — states of our sensory systems caused by the stimulation of our sense organs — or that sensory processes mediate our perceptual consciousness of the world. Sensations, on this picture, are *causes* for our perceptual experience of objects and qualities, not *direct objects* of these experiences. To say that we are phenomenally conscious of our sensations, as opposed to the objects those sensations make available to us, is, as the fourteenth–fifteenth-century Tibetan philosopher Tsongkhapa (2006) puts it, to confuse the epistemic instrument with the epistemic object. When we see the moons of Jupiter through a telescope, we see the moons of Jupiter, not light refracted through lenses. When we focus on the light refracted through the lenses, we take a higher-order perspective on that perceptual process, but that objectifies the light; it does not create experience of subjectivity.

While it makes perfectly good sense to refer to *blue sensations*, that is only to say that those sensations are the kinds of events that mediate the perception of blueness, not that they are blue in the sense that blue objects are, let alone in some other mysterious, private sense. No sensation has any perceptible properties — when we attend carefully to our *experience*, we attend carefully to the *world we experience*, not to an inner life whose properties are known only to us, but whose qualities we magically learn to report.

5. Non-duality and the illusion of givenness

The Buddhist philosopher Vasubandhu (fourth–fifth century CE) put the same point in terms of his analysis of experience in terms of the doctrine of the three natures that constitutes the heart of Yogācāra phenomenology. Vasubandhu, in his 'Treatise on the Three Natures' (Garfield, 2009; 2015) argues that every object of experience has three distinct but interdependent natures (*svabhāva*): each has an imagined (*parikalpita*) nature — that we take it to have in virtue of a combination of our prejudices and innate cognitive reflexes; each has a dependent (*paratantra*) nature, representing its underlying causal structure; and each has a consummate (*parinispanna*) nature — the nature we see that it has when we empty the dependent of the imagined. (See Garfield, 2009; 2015, pp. 186–9; Gold, 2006, for a detailed exposition.)

So far, we have been taking for granted a naïve bifurcation of experience into a subjective and an objective aspect. Vasubandhu calls that understanding of our experience into question, arguing that this duality is in fact illusory. He is not arguing for some kind of mystical dissolution into a formless void, but rather calling attention to the role of cognition in constructing our sense of who we are, and of the fact that that construction is often mistaken for a simple taking of what is given. He deploys the three nature theory to make this point.

The imagined nature is understood to involve the projection of subject–object duality and a distinction between an external world of objects and an inner world of experiences. Vasubandhu argues that we naïvely take the objects of our experience to exist externally to us in exactly the way we perceive them, taking perceived blue and perceived C# to look exactly as our eyes and ears deliver them, and to take our sensory systems as transparent windows onto a world existing with sensible properties independent of our mode of apprehension.

The other side of this is that we take our own subjectivity in this experience to consist in the inner direct experience of the qualities of these objects, and distinguish the immediately apprehended experience from the objects experienced, issuing in the dual-aspect structure of our representation of our experience. This construction of the structure of experience as involving an immediately apprehended subjective dimension and a corresponding mediated apprehension of an objective world to which it corresponds is the result of our innate response to perception.

The dependent nature, Vasubandhu argues, is just the fact that the objects of our experience are known to us through causal processes: our access to the external world is a causal interaction between distal objects, our sensory apparatus, and our cognitive mechanisms. Experience is not a separate entity that arises from this: it is this interaction, and the objects of our experience are constituted in that interaction; perceptual experience just is that interaction with the world around us.

The consummate nature of things, according to Vasubandhu, is, in his phrase, the fact that the dependent nature is empty of the imagined. That is, it is the fact that the set of causal processes in which perception consists is entirely empty of a division into subject and object or into inner and outer. There is just a causal stream constituting the causal interaction of an organism and the world. That stream is experienced as (imagined to be) the apprehension by a subject of a world, delivered exactly as it is into an inner space. But that, Vasubandhu argues, is an illusion, the illusion that the conceptual response to sensory experience represented by the imagined nature is a transparent delivery of experience as it is.

The idea that patches of blue or sounds of oboes exist independently as they are experienced by us, and then are reproduced inside us in experience, is a fantasy that sounds crazy the moment we make it explicit. So, Vasubandhu concludes, both the idea that the objects we experience exist external to us, and the idea that we have appearances internal to us, are each products of imagination. Instead, we simply causally interact with a world around us, through sensory systems to whose outputs we respond conceptually, confusing that conceptual response with immediate awareness. That is what he means when he states that the imagined nature is the imagination of subject–object duality.

The important insight for present purposes is this: the phenomenal realist argues that there is no gap between appearance and reality when it comes to experience itself; that our inner life is given to it just as it is, with its own phenomenal properties distinct from those of external objects; that just as there is something that external objects are like, there is something that our inner experience is like. This requires a firm duality between the objective and the subjective, with a distinct set of inner phenomena that are like something. Vasubandhu calls that into question, arguing that it is an imaginary superimposition on a reality that has no such structure. On this view, the very bifurcation of experience into the subjective and the objective presupposed by the realist about phenomenal consciousness is illusory, and the entire framework in which we understand our inner experience is subject to massive illusion.

This Indian Yogācāra Buddhist analysis presents yet another perspective on why this kind of phenomenal realism represents yet another form of the Myth of the Given, and on why that myth is indeed a myth. The Myth of the Given is the myth that there is some level of our experience that is immediate, immune from error, given to us, as opposed to constructed, and that that level of experience constitutes the foundation or transcendental condition of the possibility of knowledge of anything else. Classical sense-data theory — Sellars' direct target — is but one form of that myth. The idea that there is an immediate level of phenomenal consciousness, a primitive sense of subjectivity, a way our inner life is, independent of how we might imagine it, is only the latest version of that myth. But the very idea that anything in our conscious life is immune from illusion, that there is anything that it is like to be a subject, *per se*, that there are inner experiences that we just have and in virtue of just having them we know immediately, is itself but one more cognitive illusion.

Acknowledgments

I thank Keith Frankish and two anonymous reviewers for helpful comments on an earlier draft of this essay.

References

Garfield, J. (1996) Casting out demons and exorcising zombies: Exposing neo-Cartesian myths in Frank Jackson's philosophy of mind, in Dowe, P., Nicholls, M. & Shotton, L. (eds.) *Australian Philosophers*, pp. 55–96, Hobart: Pyrrho Press.

Garfield, J. (2009) Vasubandhu's *Trisvabhāvanirdeśa*, in Edelglass, W. & Garfield, J. (eds.) *Buddhist Philosophy: Essential Readings*, New York: Oxford University Press.

Garfield, J. (2015) *Engaging Buddhism: Why it Matters to Philosophy*, New York: Oxford University Press.

Gold, J. (2006) No outside, no inside: Duality, reality and Vasubandhu's illusory elephant, *Asian Philosophy*, **33** (2), pp. 1–38.

Kriegel, U. (2007) The phenomenologically manifest, *Phenomenology and the Cognitive Sciences*, **6**, pp. 115–136.

Kriegel, U. (2009) *Subjective Consciousness: A Self-Representational Theory*, New York: Oxford University Press.

Nagel, T. (1974) What is it like to be a bat, *Philosophical Review*, **83**, pp. 435–450.

Sellars, W. (1963) Empiricism and the philosophy of mind, in *Science, Perception and Reality*, London: Routledge and Kegan Paul.

Tsongkhapa (2006) *Ocean of Reasoning: A Great Commentary on Nāgarjuna's Mūlamadhyamakakārikā*, Samten, N. & Garfield, J. (trans.), New York: Oxford University Press.

Wittgenstein, L. (1953) *Philosophical Investigations*, New York: Macmillan.

Zahavi, D. (2005) *Subjectivity and Selfhood: Investigating the First-Person Perspective*, Cambridge, MA: MIT Press.

Philip Goff

Is Realism about Consciousness Compatible with a Scientifically Respectable Worldview?

Abstract: *Frankish's argument for illusionism — the view that there are no real instances of phenomenal consciousness — depends on the claim that phenomenal consciousness is an 'anomalous phenomenon', at odds with our scientific picture of the world. I distinguish two senses in which a phenomenon might be 'anomalous': (i) its reality is inconsistent with what science gives us reason to believe, (ii) its reality adds to what science gives us reason to believe. I then argue (A) that phenomenal consciousness is not anomalous in the first sense, and (B) the fact that phenomenal consciousness is plausibly anomalous in the second sense is only problematic if it can be shown that our introspectively-based reasons for believing in consciousness are epistemically problematic. I finish by suggesting that Frankish might be motivated to adopt radical naturalism because he takes doing so to be the appropriate response to the incredible success of natural science. I outline a way of thinking about the history of science which undermines this motivation.*

Illusionism is the view that the apparent reality of phenomenal consciousness is a powerful illusion; illusionists deny that anything really has phenomenal consciousness. This is a position which deserves exploration, and Keith Frankish's recent defence of it is rigorous and compelling. It also brings together, perhaps for the first time, the varied literature on this topic.

Correspondence:
Email: goffp@ceu.edu

Many philosophers will protest that illusionism is obviously false, or that the assertion of it is somehow self-defeating. I am inclined to think both of these things. However, my aim in this piece is not to argue that illusionism cannot be true, but rather to undermine the motivation for it. Frankish believes that realism about phenomenal consciousness is at odds with scientifically respectable metaphysics. I hope to show that this is not the case, or at least that Frankish has given us no grounds for thinking that it is.

1. Is illusionism coherent?

Phenomenal consciousness is typically defined as follows:

> Something is phenomenally conscious if and only if there is something that it's like to be it.

In his paper Frankish articulates a deflationary understanding of 'what it's like' talk, and hence for the sake of clarity we can add that phenomenal consciousness involves *experience of qualities*: the redness of red experiences, the bitter quality involved in the experience of biting into a lemon, the qualitative character of heat sensations. Indeed it is precisely the qualitative character of consciousness that Frankish finds problematic. In what follows I will use the word 'consciousness' to refer to phenomenal consciousness.

The illusionist accepts that human beings *represent* states of consciousness, and indeed that we typically *believe* that states of consciousness exist. But she also holds that all such representations are non-veridical, and the corresponding beliefs false. Just like beliefs about the Loch Ness Monster, beliefs about consciousness fail to correspond to anything real.

Beliefs are a kind of propositional thought. Is it coherent to accept the reality of thought whilst denying the reality of consciousness? That depends on whether or not there is a constitutive relationship between thought and consciousness. Frankish assumes throughout the paper that we can account for thoughts, such as beliefs and other mental representations, without the postulation of consciousness. In this he follows the dominant view in analytic philosophy that there is no essential connection between thought and consciousness. This view was largely unquestioned in the twentieth century.[1] However, there is now a growing movement in analytic philosophy defending the thesis that thoughts, and indeed mental representations in general, are identical with (or directly constituted of) forms of phenomenal consciousness. Uriah Kriegel has dubbed this movement 'the Phenomenal Intentionality Research Programme' (Kriegel, 2013).

Clearly, if the convictions of the Phenomenal Intentionality Research Programme turn out to be correct, then illusionism involves a straightforward contradiction: you can't assert the existence of thought but deny the existence of consciousness if thought just is a (highly evolved) form of consciousness. I believe there is strong reason to accept that thought is a form of consciousness, and hence strong reason to think that illusionism is indeed incoherent. However, I will not have space to defend this here. I merely note that illusionism depends on an assumption which Frankish has not defended in this paper: that mental representation is not dependent on phenomenal consciousness. In what follows I will grant him this assumption for the sake of argument.

2. Is consciousness anomalous?

Frankish's principal motivation for illusionism is his conviction that consciousness is an 'anomalous' phenomenon, at odds with our scientific picture of reality. It might be useful to distinguish two ways in which a certain thesis might be in tension with our scientific knowledge of the world:

> *Being Anomalous by Inconsistency* — A thesis is anomalous by inconsistency iff it is inconsistent with what we know (or have good to reason to believe) about reality through empirical methods. An example of this is the creationist thesis that the world is between 6,000 and 10,000 years old.

> *Being Anomalous by Irreducibility* — A thesis is anomalous by irreducibility iff it requires ontological commitments over and above those required by what we know (or have good reason to believe) about reality through empirical methods. A plausible

---

[1] There were a few opponents this consensus, e.g. John Searle (1984), Galen Strawson (1994), and Charles Siewert (1998).

example of this is the thesis that there is a non-interventionist God.[2]

At various points Frankish seems to suggest that consciousness is anomalous in both of these senses. In the next section I will argue that the reality of consciousness is not anomalous by inconsistency. I will then go on to argue that, although there may be grounds for thinking that consciousness may be anomalous by irreducibility, Frankish has given us no reason to think that this fact is in any way problematic for the consciousness realist.

2.1. Is phenomenal consciousness anomalous by inconsistency?

Why might one think that the reality of consciousness is inconsistent with what we know about reality scientifically? Frankish supports the standard arguments of Chalmers, Jackson, and others to the conclusion that states of consciousness are non-physical.[3] This in itself does not entail an inconsistency with what we know empirically; the postulation of non-physical properties may merely add to what we know about the world empirically, without contradicting it. However, inconsistency may arise when we try to find a causal role for such non-physical consciousness in the material world.

Paul Churchland, among others, has argued that giving a causal role to non-physical properties violates principles of energy conservation:

> ...[A]s has been known for more than fifty years... forms of Dualism... fly in the face of basic physics itself... since any position that includes non-physical elements in the causal dynamics of the brain must violate both the law that energy is neither created nor destroyed, and the law that the total momentum in any closed system is always conserved. In short, you simply can't get a change in any aspect of the physical brain (for that would causally require both energy changes and momentum changes) save by a compensatory change in some other physical aspect of the brain, which will thereby lay claim to being the cause at issue. There is simply no room in a physical system for ghosts of any kind to intervene in some fashion to change its dynamical behavior. Any physical system is 'dynamically closed' under the laws of Physics. (Indeed, it was this very difficulty, over a century ago, that initially

---

[2] This thesis is anomalous by irreducibility only if design arguments for the existence of God fail.

[3] Chalmers (1996; 2009), Jackson (1982; 1986). I critique these forms of the argument and suggest an alternative in Goff (2011) and Goff (forthcoming).

motivated the desperate invention of Epiphenomenalism in the first place.) (Churchland, 2014)

This argument is far too hasty. As David Papineau has pointed out, fundamental mental forces need not constitute a counter-example to the principle of the conservation of energy, so long as those forces act conservatively; so long as they 'operate in such a way as to "pay back" all the energy they "borrow" and vice-versa':

> ...[T]he conservation of energy in itself does not tell which basic forces operate in the physical universe. Are gravity and impact the only basic forces? What about electro-magnetism? Nuclear forces? And so on. Clearly the conservation of energy as such leaves it open exactly which basic forces exist. It requires only that, whatever they are, they operate deterministically and conservatively. (Papineau, 2002, p. 252)

We must distinguish this kind of argument, based on conservation of energy, from more common and more promising arguments based on the alleged causal closure of the physical. Frankish's concerns about the causal role of phenomenal consciousness are of the latter kind:

> Non-physical properties can have no effects in a world that is closed under causation, as ours appears to be, and the mind sciences show no independent need to refer to exotic physical processes, such as quantum-mechanical ones. The threat of epiphenomenalism hangs over radical theories. (Frankish, this issue, p. 25)

This familiar worry is perhaps most associated with Jaegwon Kim (1989). If we grant the premise that consciousness is non-physical, then we can turn this worry into an argument that non-epiphenomenal consciousness is anomalous by contradiction:

1. The thesis of causal closure (i.e. the thesis that every physical event has a sufficient physical cause) is empirically well supported.
2. Consciousness is non-physical.
3. If consciousness exists, and it is not epiphenomenal (i.e. it has a causal impact on the physical world), then there would be physical events (i.e. the events caused by consciousness) which have a non-physical rather than a physical cause, in violation of causal closure.[4]

---

[4] The truth of this premise relies on an implicit assumption that the effects of consciousness are not systematically over-determined. This premise is usually explicit in discussions of this issue, but I leave it implicit here in order to keep things simple.

4. Therefore, either consciousness is epiphenomenal or its existence is inconsistent with an empirically well-supported thesis (i.e. causal closure).

The thesis that the physical world is causally closed is often stated, but very rarely defended. Following this trend Frankish does not defend, nor reference any defence of, the principle, beyond saying that it 'appears to be true'. Peter Rauschenberger (ms) has recently given an extensive critique of the scant empirical defence of causal closure which has appeared in the literature.[5]

Moreover, as Frankish points out, even if causal closure is true, there is a way of reconciling the non-physicality of consciousness with the causal closure of the physical world: Russellian monism.[6] Russellian monism results from a recent rediscovery of the approach to the mind–body problem defended (independently) by Russell and Eddington in the 1920s (Russell, 1927; Eddington, 1928). The starting point for the view is Russell's observation that the physical sciences, such as physics, chemistry, and neuroscience, characterize physical properties *dispositionally*, i.e. solely in terms of what those properties *do*. In neuroscience, for example, a given brain state is characterized in terms of (i) its causal role in the brain, (ii) its chemical constituents. Those chemical constituents are then characterized in terms of (i) their causal role, and (ii) their physical constituents. And the basic physical constituents are characterized entirely dispostionally; mass, for instance, is characterized in terms of gravitational attraction and the disposition to resist acceleration.

Bearing this in mind, consider a case in which a certain brain state is strongly correlated with a certain conscious state. Suppose, for the sake of having an example, that we discover a strong correlation between hunger and lateral hypothalamus stimulation (LHS). Physical science tells us nothing about what LHS *is* beyond what it *does* (and what its constituents do). Now there are powerful arguments to the conclusion that there must be more to the nature of a property than its causal role, and if those arguments are successful then there must be more to LHS than what physical science tells us about it (Robinson, 1982; Blackburn, 1990; Goff, forthcoming, chapter 6). But even if those arguments don't work, it is surely coherent to suppose that there

---

[5] The appendix of Papineau (2002) is perhaps the best defence of causal closure, although even here the empirical support is explored in very little detail.

[6] For a recent collection of essays on Russellian monism, see Alter and Nagasawa (2015).

is more to LHS that its causal role, that LHS has a categorical nature which underlies its dispositional characteristics. If Frankish thinks that the rather orthodox philosophical view that dispositional properties are grounded in categorical properties is incoherent or implausible, then he owes us an argument to justify that conclusion.

The Russellian monist has a proposal for what the categorical nature of LHS is (at least in part): the feeling of hunger. According to Russellian monism, physical science describes LHS 'from the outside', i.e. it tells us what it does; whilst in introspection we know LHS 'from the inside' as the conscious experience of hunger. LHS is a single state with a qualitative intrinsic nature which underlies the dispositional characteristics which are the focus of empirical science.

Now, if hunger is a non-physical state, and hunger is the intrinsic nature of LHS, does it follow that LHS is a non-physical state? In fact, in the context of Russellian monism we need to note certain ambiguities in the term 'physical property'. If Russellian monism is true, then LHS is *referred to* by the predicates of neuroscience, and in that sense LHS is 'physical'; but it has a categorical nature which is not revealed by physical science, and in that sense it is 'non-physical'.[7]

It is in the latter sense that the standard 'anti-physicalist arguments', such as the zombie conceivability argument and the knowledge argument, try to demonstrate that consciousness is 'non-physical'. But consciousness only needs to be 'physical' in the former sense in order to secure its causal role. If the conscious feeling of hunger is the categorical nature of lateral hypothalamus stimulation, then it is in virtue of being so that it is ensured a genuine causal role in the material world.

Frankish objects to Russellian monism on the grounds that 'it involves huge profligacy with phenomenal properties and preserves the potency of consciousness only at the cost of making all physical causation phenomenal' (p. 25). However, this objection assumes that Russellian monism entails panpsychism, which is not the case. Although there are panpsychist versions of Russellian monism, such as the view of Strawson which Frankish references, there are also non-panpsychist versions.[8] It is perfectly possible to be a Russellian monist

---

[7] Stoljar (2001) outlines a detailed account of these two definitions of physicality.

[8] For example, Pereboom (2011) defends a disjunction of non-panpsychist Russellian monism and illusionism. I favour a panpsychist version of Russellian monism (Goff, forthcoming), but those impressed by Frankish's worries need not follow me in this.

whilst only believing in those forms of consciousness we have a pre-theoretical commitment to.

There is therefore no reason to suppose that there is any tension between full-blooded realism about non-epiphenomenal qualitative phenomenal consciousness and our empirical knowledge of the world; there are no grounds for thinking that non-epiphenomenal consciousness is anomalous by contradiction. Even if we have reason to believe that the physical world is causally closed, and this has not been adequately defended, there is a way of reconciling causal closure with a commitment to conscious states which are both causally efficacious and non-physical ('non-physical' in the sense of having a nature not fully captured by physical science, and this is all that is demonstrated by the standard 'anti-physicalist' arguments).

2.2. Is it a problem if phenomenal consciousness is anomalous by irreducibility?

Russellian monism allows us to accommodate full-blooded qualitative consciousness into the physical world without contradicting anything we know about the world empirically. But it does involve *adding* to what we know about the world empirically (I am here taking 'empirical' knowledge to be knowledge which results from third-person observation and experiment). Both the Russellian monist and the illusionist accept that states of phenomenal consciousness are not accessible to third-person scientific methods, and are not explicable in terms of the dispositional properties that feature in physical science. If this is correct, then realism about consciousness takes us beyond the metaphysical commitments of third-person science.

Frankish assumes throughout the paper that it is problematic to go beyond the metaphysical commitments of third-person science. But why should we think this? It goes without saying that we should not believe in things that we have no reason to believe in. But the realist would claim that our apparent introspective awareness of our own consciousness *does* give us reason to believe in the reality of consciousness. If we have reason to believe in consciousness, and if there is good reason to believe (as Frankish thinks there is) that consciousness cannot be explained in terms of the postulations of third-person science, it follows that we have reason to believe in something over and above the postulations of third-person science.

Of course, if there were a tension between the reality of consciousness and the facts of natural science, then we would have to do some weighing of epistemic reasons. However, we demonstrated in the last section that there is no such tension. Once we have dispelled the worry that consciousness is anomalous by contradiction, the onus is on the illusionist to demonstrate that our apparent introspective awareness of consciousness gives us no reason to believe in the reality of consciousness. Frankish accepts that it is epistemically permissible to believe in the external world on the basis of our senses. Why is it not epistemically permissible to believe in phenomenal consciousness on the basis of introspection?

Frankish points to empirical evidence that first-person judgments are not reliable. But it is a huge leap to go from the fact that our putative introspective access to phenomenal consciousness is not perfect, to the claim that it has no epistemic force at all. If there is no empirical case *against* the reality of consciousness, all the realist needs to hold is that our apparent introspective awareness of phenomenal consciousness gives us reasonable grounds to believe that phenomenal consciousness exists. Frankish is obliged to tell us why this belief is unreasonable.

There are of course epistemic disadvantages pertaining to our epistemic access to consciousness. Facts about phenomenal consciousness cannot be intersubjectively checked in the way that facts about the external world can. This is unfortunate, and introduces all sorts of methodological difficulties to the science of consciousness. However, I can't see an argument from the fact that our apparent access to phenomenal consciousness is non-ideal to the conclusion that it ought not to be trusted at all. Third-person access to the external world is non-ideal too; for example, there seems to be no way of purging it of sceptical doubt (arguably our access to consciousness does not suffer from this drawback, but perhaps it would be question-begging to assert that in this context). The epistemic situation of evolved creatures is significantly inferior to the epistemic situation of angels; we've got to take what we can get.

The most interesting arguments Frankish raises against the plausibility of consciousness realism focus on the peculiar epistemic relationship that seems to obtain between the mind and its conscious states. I will spend the rest of this section focusing on these arguments in some detail.

Frankish claims that, 'If realists are to maintain that phenomenal consciousness is a datum, then they must say that we have a special kind of epistemic access to it, which excludes any possibility of error' (p. 30). I am not clear why he thinks that consciousness must be known with certainty in order to be a datum; the realist might merely claim that introspection provides good reason to believe in the reality of consciousness, just as the senses provide good reason to believe in the external world. Still, it is true that many realists, especially those who take consciousness to be irreducible, do believe that conscious subjects stand in a special, non-causal relationship of 'acquaintance' to their conscious states, and that in virtue of the relationship of acquaintance the existence and nature of conscious states is known with something close to certainty (Chalmers, 2003; Goff, 2015; forthcoming).

Frankish raises a few worries about the relationship of acquaintance:

> First, acquaintance can have no psychological significance. In order to talk or think about our phenomenal properties, we need to form mental representations of them, and since representational processes are potentially fallible, the certainty conferred by acquaintance could never be communicated... (p. 30)

In response to this worry we can note that if there is such a thing as acquaintance, then it is *itself* a kind of representational process: in being acquainted with pain I represent that property to myself. And no consciousness realist that I know of holds that *all* representations of consciousness are certain or infallible; David Chalmers for example holds that only a narrow class of mental concepts — those he calls 'direct phenomenal concepts' — have this special status. Thus, I can't see why any realist would worry that the certainty involved in acquaintance-based representations of consciousness is lost when the content of the concept is incorporated into other representational processes. Again Frankish seems to be assuming that the reality of phenomenal consciousness is a datum only if it is known with certainty, and I fail to see what reason we have to think that. My experience of the table in front of me does not guarantee the table's existence, but it is nonetheless reasonable to believe in the table on the basis of it.

> Second, acquaintance theory assumes that reactions and associations a sensory episode evokes do not affect its feel, since we are not directly acquainted with them or their effects. Yet there is reason to think that our reactions and associations do shape our sense of what our experiences are like. (p. 31)

It would be good to hear more detail about how the evidence Frankish references counts against an acquaintance model. For the moment I am not clear why the acquaintance theorist cannot accept that certain associations and tendencies to react do have a causal impact on phenomenal states, phenomenal states which are subsequently known in acquaintance-based representations. And given that the acquaintance theorist holds that only a limited subset of representations of consciousness involve certainty, she is quite entitled to hold that reactions and associations cause a certain amount of misrepresentation among our representations of consciousness. Perhaps there are some empirical worries about acquaintance, but I think we need to hear more detail.

> Acquaintance theory also comes with heavy metaphysical baggage. It is hard to see how physical properties could directly reveal themselves to us... Moreover, it may require an anti-physicalist view of the experiencing subject too. If subjects are complex physical organisms, how can they become directly acquainted with consciousness? (p. 31)

This seems to be an Ockhamist objection to realism about direct acquaintance relations and subjects which stand in such relations. But if we have reason to believe in consciousness, and we have reason to believe that consciousness involves subjects bearing direct acquaintance relations, then we have reason to believe in subjects bearing direct acquaintance relations. If there was a clash between acquaintance and our empirical knowledge of person and brain, then we might have grounds for rejecting acquaintance. But cognitive science and neuroscience are limited to tracking the dispositional features of brains. It could be that the brain also has a categorical nature, and that that categorical nature involves a subject bearing direct acquaintance relations. If we have reason to believe in consciousness, and this is what consciousness involves, then this is what we ought to believe (in the absence of a powerful counter-reason).

To summarize my argument so far:

- Frankish has given us no reason to believe that phenomenal consciousness is anomalous by inconsistency. Even if the physical world is causally closed, and this has not been adequately defended, Russellian monism provides a way of reconciling causal closure with full-blooded realism about phenomenal consciousness.

- If the conceivability argument or the knowledge argument is sound, then realism about phenomenal consciousness requires metaphysical commitments over and above those of third-person observation and experiment. However, if we have reason to believe in phenomenal consciousness, then we have epistemically legitimate grounds for making those additional commitments. And Frankish has given us no good reason to doubt that our apparent introspective access to phenomenal consciousness gives us good grounds for believing in the reality of phenomenal consciousness (just as our apparent sensory access to the external world gives us good grounds for believing in the external world).

3. Radical naturalism

Why is Frankish so convinced that we have no good reason to believe in consciousness? I suspect that the fundamental motivation for Frankish, and for the other illusionists he discusses, is a deep commitment to a radical form of methodological naturalism, which we can define as follows:

> *Radical naturalism* — Third-person observation and experiment, and only third-person observation and experiment, should be our guide in finding out the nature of reality.

The reality of consciousness does not seem to be accessible to third-person observation and experiment; I can't see your pain no matter how much I poke around in your head. And thus, if we accept radical naturalism, illusionism becomes extremely plausible. The question we now need to ask is: what reason do we have to be radical naturalists?

I believe that the attraction to radical naturalism arises from an emotional response to the success of science. It cannot be denied that in the last five hundred years or so the project of natural science has gone extremely well. From the movement of the planets, to the evolution of life, to the fundamental constituents of matter, natural science seems to be an unstoppable juggernaut of explanation. For the radical naturalist, what this shows is that we've finally found something that *works*, something we can put our ontological faith in. For thousands of years before the scientific revolution philosophers struggled to find out what reality is like and got nowhere. Since the scientific revolution natural science has enjoyed success after success after success. From this perspective, philosophers who look somewhere other than third-person empirical science to try to work out what reality is like are 'old school', dragging us back to the Dark Ages. They are to be equated with folk who believe in magic, or deny climate change, or think that the world was created in six days.

However, there is a deep irony here. A key moment in the scientific revolution was Galileo's declaration that mathematics should be the language of natural science. But Galileo felt able to make this declaration only once he had stripped the material world of *sensory qualities*. For Galileo, colours, tastes, sounds, and odours aren't really in *matter*, but in the *experiencing soul* of the perceiver.[9] By stripping matter of its qualities, leaving only size, shape, location, and motion, Galileo gave us, for the first time in history, a material universe which could be exhaustively described in the austere language of mathematics (Galileo, 1623/1957, pp. 274–7).

Thus, natural science begins with Galileo *limiting its scope of enquiry*, by putting the sensory qualities in the soul and entreating 'natural philosophers' to focus only on what can be captured mathematically. This limited project has gone extremely well, allowing the construction of mathematical models which yield extremely accurate predictions of the behaviour of matter. This in turn has enabled us to manipulate the natural world in extraordinary ways, and thus to produce technology undreamt of by previous generations. This incredible success has created in contemporary philosophers a great desire to place all of their ontological faith in natural science.

However, it is clear upon calm reflection that the fact that *things go really well when we ignore the sensory qualities* gives us absolutely no reason to think that *sensory qualities don't exist*. It would be nice if we could apply the methods of third-person science to the qualities of experience, but their private nature is incompatible with public observation, and (as Galileo realized, and Frankish concurs) their qualitative nature resists (exhaustive) mathematical characterization. This doesn't mean they don't exist; it just means that the human epistemic situation is far from ideal. We have at least as much reason to believe in the qualities of experience — on the basis of introspection — as we do to believe in the external world — on the basis of perception. And if we want a complete theory of the world, then we must try to construct one which incorporates everything we have reason to believe in, not just those things which are easier to deal with.

Natural science has done an extremely good job of describing the causal structure of matter. The job of the metaphysician is to build on this, by developing a theory which unifies the findings of natural science with other things we have reason to believe in. What other

---

[9] In contrast to Descartes, Galileo conceived of the soul in Aristotelian terms, as the principle of animation in the body (*corpo sensitivo*). See Ben-Yami (2015, chapter 3) for more discussion of this.

things do we have reason to believe in? At the very least the experiential qualities which Galileo set outside of the scope of natural science.

The project of metaphysics is currently hampered by an understandable but irrational attraction to scientism, which results from the visceral impact of technology on philosophical inclinations. So much so that the general public has almost no involvement with the metaphysics which goes on in philosophy departments, nor even awareness that it is going on. At some point this irrational attraction to scientism will fade, and society as a whole will return to the noble project of trying to work out what reality is like; this time armed with rich information about the causal structure of reality, information which was not available to our ancestors.

Many bemoan the fact that metaphysics doesn't seem to have got anywhere. I'm inclined to think it hasn't really begun.

References

Alter, T. & Nagasawa, N. (eds.) (2015) *Consciousness and the Physical World*, New York: Oxford University Press.
Ben-Yami, H. (2015) *Descartes' Philosophical Revolution: A Reassessment*, Basingstoke: Palgrave Macmillan.
Blackburn, S. (1990) Filling in space, *Analysis*, **50**, pp. 62–65.
Chalmers, D.J. (1996) *The Conscious Mind: In Search of a Fundamental Theory*, New York: Oxford University Press.
Chalmers, D.J. (2003) The content and epistemology of phenomenal belief, in Smith, Q. & Jokic, A. (eds.) *Consciousness: New Philosophical Perspectives*, Oxford: Oxford University Press.
Chalmers, D.J. (2009) The two-dimensional argument against materialism, in McLaughlin, B. (ed.) *Oxford Handbook of the Philosophy of Mind*, pp. 313–339, Oxford: Oxford University Press.
Churchland, P. (2014) Consciousness and the introspection of phenomenal simples, in Brown, R. (ed.) *Consciousness Inside and Out: Phenomenology, Neuroscience, and the Nature of Experience*, vol. 6 of Studies in Brain and Mind, Dordrecht: Springer.
Eddington, A. (1928) *The Nature of the Physical World*, Cambridge: Cambridge University Press.
Galileo Galilei (1623/1957) *The Assayer*; reprinted in Drake, S. (ed.) *Discoveries and Opinions of Galileo*, New York: Doubleday.
Goff, P. (2011) A posteriori physicalists get our phenomenal concepts wrong, *Australasian Journal of Philosophy*, **89** (2), pp. 191–209.
Goff, P. (2015) Real acquaintance and physicalism, in Coates, P. & Coleman, S. (eds.) *Phenomenal Qualities: Sense, Perception, and Consciousness*, Oxford: Oxford University Press.
Goff, P. (forthcoming) *Consciousness and Fundamental Reality*, Oxford: Oxford University Press.

Jackson, F. (1982) Epiphenomenal qualia, *Philosophical Quarterly*, **32**, pp. 127–136.
Jackson, F. (1986) What Mary didn't know, *Journal of Philosophy*, **83**, pp. 291–295.
Kim, J. (1989) Mechanism, purpose, and explanatory exclusion, *Philosophical Perspectives*, **3**, pp. 77–108.
Kriegel, U. (ed.) (2013) *Phenomenal Intentionality*, New York: Oxford University Press.
Papineau, D. (2002) *Thinking about Consciousness*, Oxford: Clarendon Press.
Pereboom, D. (2011) *Consciousness and the Prospects of Physicalism*, New York: Oxford University Press.
Rauschenberger, P. (ms) The dogma of causal closure, [Online], https://independent.academia.edu/P%C3%A9terRauschenberger.
Robinson, H. (1982) *Matter and Sense*, Cambridge: Cambridge University Press.
Russell, B. (1927) *The Analysis of Matter*, London: Kegan Paul.
Searle, J. (1984) *Mind, Brains and Science*, Cambridge, MA: Harvard University Press.
Siewert, C. (1998) *The Significance of Consciousness*, Princeton, NJ: Princeton University Press.
Stoljar, D. (2001) Two conceptions of the physical, *Philosophy and Phenomenological Research*, **62**, pp. 253–281.
Strawson, G. (1994) *Mental Reality*, Cambridge, MA: MIT Press.

Michael S.A. Graziano

Consciousness Engineered

Abstract: *The attention schema theory offers one possible account for how we claim to have consciousness. The theory begins with attention, a mechanistic method of handling data in which some signals are enhanced at the expense of other signals and are more deeply processed. In the theory, the brain does more than just use attention. It also constructs an internal model, or representation, of attention. That internal model contains incomplete, schematic information about what attention is, what the consequences of attention are, and what its own attention is doing at any moment. This 'attention schema' is used to help control attention, much like the 'body schema', the brain's internal simulation of the body, is used to help control the body. Subjective awareness — consciousness — is the caricature of attention depicted by that internal model. This article summarizes the theory and discusses its relationship to the approach to consciousness that is called 'illusionism'.*

1. Introduction

Recently my colleagues and I proposed the attention schema theory as an explanation of consciousness (Graziano, 2013; 2014; Graziano and Kastner, 2011; Kelly *et al.*, 2014; Webb and Graziano, 2015; Webb, Kean and Graziano, 2016). Here by 'consciousness' I mean that, in addition to processing information, people report that they have a conscious, subjective experience of at least some of that information. The attention schema theory is a specific explanation for how we make that claim. It is a theory of how information is constructed in the brain and used to model the world and guide decisions, conclusions, speech, and behaviour. It is a theory of how the human machine claims to have consciousness and assigns a high degree of certainty to that conclusion.

Correspondence:
Michael S.A. Graziano, Dept. of Psychology, Princeton Neuroscience Institute, Princeton University, Princeton, NJ 08544. *Email: graziano@princeton.edu*

2. Build-a-brain

One useful way to introduce the theory is through the hypothetical challenge of building a robot that asserts it is subjectively aware of an object and describes its awareness in the same ways that we do. I argue that the construction outlined below is not simply an academic exercise in engineering a zombie. Instead that type of mechanism is so basic that it is likely to have evolved in the brain. Moreover, as discussed in the second half of the article, growing evidence suggests that something like that mechanism *does* exist in the brain.

Figure 1 shows a robot looking at an apple. What information should be incorporated into its brain? First, we give it information about the apple (Figure 1A). Light enters the eye, is transduced into signals, and the information is processed to construct a description of the apple that includes shape, colour, size, location, and other attributes. This representation, or internal model, is constantly updated as new signals arrive. The model is schematic. It is a simplified proxy for the real thing. Given the limited processing resources in the brain, internal models are necessarily incomplete and simplified. They are efficient. They are data-compressed. Here we give our robot just such a simplified, schematic internal model of an apple.

Is the robot in Figure 1A aware of the apple? In one sense, yes. The term 'objective awareness' is sometimes used to indicate that the information has gotten in and is being processed (e.g. Szczepanowski and Pessoa, 2007). The machine in Figure 1A is objectively aware of the apple. But does it have a *subjective* experience?

To help explore that question we add a user interface, the linguistic processor shown in Figure 1B. Like a search engine, it can take in a question, search the internal model, and answer. We ask, 'What's there?' It answers, 'An apple'. We ask, 'What are the properties of the apple?' It answers, 'It's red, it's round, it's at that location'. It can provide those answers because it contains that information.

Figure 1B could represent an entire category of theory about consciousness, such as the global workspace theory (Baars, 1988; Newman and Baars, 1993). In that theory, consciousness occurs when information is broadcast globally throughout the brain. In Figure 1B, the sensory representation of the apple is broadcast globally, and as a result the cognitive and linguistic machinery has access to information about the apple. The robot can therefore report that the apple is present.

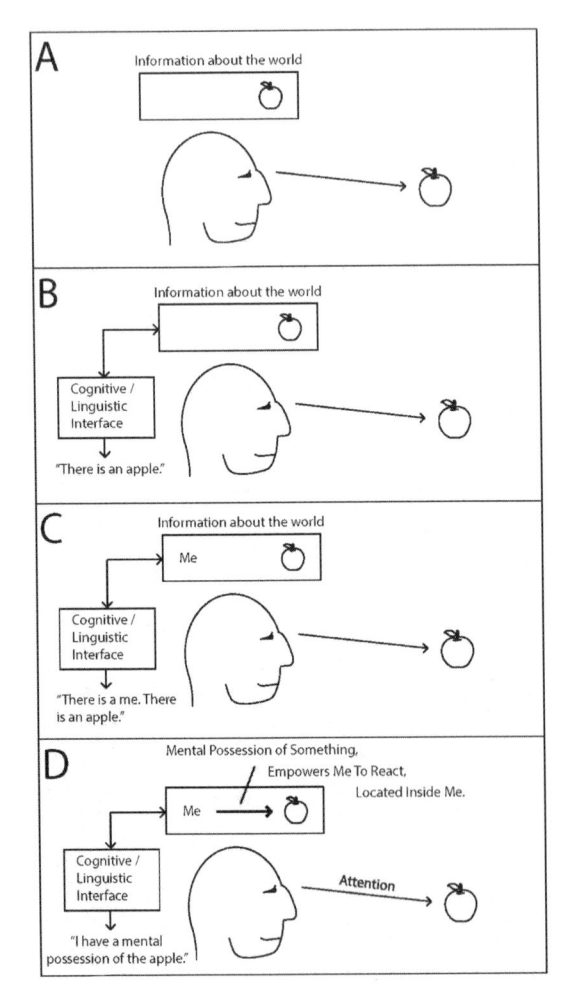

Figure 1. Construction of the attention schema theory. A robot has information about the world in the form of internal models. A. The robot has an internal model of the apple. B. The robot has a linguistic interface that acts as a search engine. It takes in questions, searches the internal model, and replies to the questions. C. The robot has a second internal model, a model of the self. D. The main components of the attention schema theory. The robot has an internal model of the self, a model of the apple, and a model of the attentional relationship between the self and the apple. That attention schema describes something physically incoherent, a caricature of attention, subjective awareness. The machine insists that it has subjective awareness of the apple because it is captive to the incomplete information in its internal models.

But Figure 1B remains an incomplete account of how a machine claims to be conscious of an apple. Consider asking, 'Are you aware of the apple?' The search engine searches the internal model and finds no answer. It finds information about an apple, but no information about what 'awareness' is or whether it has any of it, and no information about what the quantity 'you' is. It cannot answer the question. It does not compute in that domain.

Perhaps we can improve the machine. In Figure 1C, a second internal model is added, a model of the self. This new internal model, like the model of the apple, is a constantly updated set of information. It might include the body schema, the brain's model of the physical self and how it moves. The self model might also include autobiographical memory and general information about personality, beliefs, and goals. If we ask the robot in Figure 1C, 'Tell us about yourself?' it can now answer. It has been given the construct of self. It might reply, 'I'm a person, I'm standing right here, I'm so tall, so wide, I can move, I grew up in Buffalo, I'm a nice guy', and so on, as its cognitive search engine accesses its internal models. Figure 1C could represent an entire category of theory in which consciousness depends on self-knowledge or self-narrative (e.g. Gazzaniga, 1970; Nisbett and Wilson, 1977).

However, once again this account is incomplete. We can ask the machine in Figure 1C, 'What is the mental relationship between you and the apple?' The search engine accesses the two available internal models and finds no answer. It finds plenty of information about the self and plenty of separate information about the apple, but no information about a mental relationship between them — no information about what a mental relationship is. Equipped only with the components shown in Figure 1C, the machine cannot even parse the question.

So far we have given the machine an internal model of the apple and an internal model of the self, but we have neglected a crucial third item present in this scene — a less concrete, more intangible item — the computational relationship between the self and the apple. We now give the machine an internal model of attention.

The word 'attention' has many meanings, some colloquial, some technical. For example, overt attention is the orienting of eyes and other sensors toward an important event. Here, by attention I refer to something often called covert attention, the deep processing of some select signals at the expense of other signals. Covert attention can move from item to item. You can shift that deep processing from the text in front of you, to the sounds coming from your back yard, to a memory that you've just recalled, to a math problem that you're solving in your head. Covert attention is a mechanistic neural phenomenon, a selective signal enhancement caused by competition among signals in the brain (Beck and Kastner, 2009; Desimone and Duncan, 1995). A person rarely looks at an apple in isolation. Other items are probably present: a plate, a table, a wall behind them, too much for the brain to process in depth at the same time. It has to prioritize. The apple signal wins the competition of the moment, is enhanced at the expense of other visual signals, and as a result can dominate the brain's outputs. The brain deeply processes information about the apple and is therefore primed to generate behaviour toward it or remember it. This is the attentive relationship between the self and the apple. An internal model of that attentive relationship is added to Figure 1D.

First consider what information might be contained in an internal model of attention. How would it describe attention? Presumably, like the internal model of the apple, it would describe useful, functional, *abstracted* properties of attention, not microscopic physical details. It might describe attention as a mental possession of something. It might describe attention as something that empowers oneself to react. It might describe attention as something located inside oneself, belonging to oneself, and not directly observable to the outside world. It might include many other abstracted properties of attention. But this internal model would not contain information about neurons, competing electrochemical signals, or other physical nuts and bolts that the brain has no pragmatic need to know. Like all internal models, it would be incomplete and schematic. It would be silent on the physical mechanisms of attention.

We ask the robot in Figure 1D, 'What is the mental relationship between you and the apple?' The search engine accesses its internal models and reports the available information. It says, 'I have a mental possession of the apple'. The answer is promising and we probe deeper. 'Tell us more about this mental possession. What are its physical properties?' For clarity, we also ask, 'Do you know what physical properties are?' The machine can answer 'Yes' because it has a body schema that describes the physical body and it has an internal model of an apple that describes a physical object. Reporting the information available to it, it might say (if it has a good vocabulary), 'I know what physical properties are. But my mental possession of the apple, the mental possession in-and-of-itself, has no physically describable properties. It's an essence located inside me. Like my arms and legs are physical parts of me, I also have a non-physical or *metaphysical* part of me. It's my mind taking hold of things — the colour, the shape, the location. My subjective self seizes those things.' The machine is describing covert attention, and the description sounds semi-magical only because it is vague on the details and the mechanistic basis of attention.

Because we built the robot, we know why it gives that answer. It's a machine accessing internal models. Whatever information is contained in those models it reports to be true. That information lies deeper than language, deeper than higher cognition. The machine insists it has subjective awareness because, when its internal models are searched, they return that information. Introspection will always return that answer. In the same way, it reports that the apple has a colour even though in reality the apple has a reflectance spectrum, not colour. Just as in Metzinger's ego tunnel (Metzinger, 2010), this brain is captive to the schematic information in its internal models.

The theory diagrammed in Figure 1D is different from a higher-order thought theory (Lau and Rosenthal, 2011). In that approach, consciousness occurs when the brain's cognitive machinery constructs a higher-order, cognitive representation, or an interpretation. Instead, in the attention schema theory, subjective awareness does not depend on cognitive or linguistic processing. It is not a construct of higher-order thought. The cognitive/linguistic layer in Figure 1 was added as a convenience to be able to query the machine, but it is not necessary. Suppose you are a rat with little cognitive and no linguistic capacity, thus the 'cognitive/linguistic' box in Figure 1D is missing. You still have the internal models themselves. The internal models in Figure 1D are fundamental, low-level representations necessary for survival: representations of self, of items in the world such as apples, and of the ever-present process of attention, the computational relationship between the self and everything else. These representations can guide behaviour directly, even without higher cognition. One way to think about Figure 1D is that the brain constructs an overarching internal model, a continuously updated simulation of its world. In that simulation, there is a self that is conscious of the apple. The brain constructs that simulation even if it lacks the sophistication to cogitate about it or talk about it.

One advantage of this theory of consciousness is that it can accommodate the correct range of information. The brain can focus attention on a colour, a shape, a motion, a sound, a touch, a memory, a thought, a fragment of autobiographical knowledge, an emotional state, or almost any other domain of information that is processed cortically. An attention schema is therefore applicable to that same range of information. One could replace the apple in Figure 1D with almost anything, whether a feature of the external world or a feature of one's internal cognition. The theory accounts for why we claim to have a conscious experience of colour, shape, sound, self, memory, emotion, and so on, and why, despite the diversity of information, the consciousness is somehow of the same nature in all cases. In the theory, an internal model of attention pertains to multiple kinds of information.

The logic of the theory can be summarized in four points. One, the brain constructs internal models of important objects and processes in the world. Therefore, two, the brain constructs an internal model of its own process of attention. Three, internal models are never accurate descriptions. They are incomplete and schematic, due to a trade-off between accuracy and processing resources. Therefore, four, a brain with an internal model of attention, even if that brain has a good enough linguistic and cognitive capacity to talk about it, would not report its attention in a physically accurate, detailed, or mechanistic manner. Instead it would claim to have something physically incoherent: a subjective mental experience. A five-word summary of the theory, that admittedly loses some nuance, is this: awareness is an attention schema.

3. Adaptive uses of the attention schema

It is clear why an internal model of an apple is useful — to guide behaviour with respect to the apple. It is also clear why an internal model of the self is useful — to monitor and thus better control one's behaviour. But what is the adaptive value of an attention schema? In the following sections I describe three uses for an attention schema, beginning with its possible role in the widespread integration of information.

3.1. Integration of information

The idea that awareness is related to the integration of information around the brain has been suggested in many forms (e.g. Baars, 1988; Crick and Koch, 1990; Damasio, 1990; Engel and Singer, 2001; Lamme, 2006; Newman and Baars, 1993; Tononi, 2008). The attention schema theory is, in its own way, a theory about the integration of information. In Figure 1D, the attention schema is a chunk of information, a descriptive model, that is linked to many disparate kinds of information. Information about the self and information about an apple are linked together by way of an intermediate bridge, the attention schema.

One can think of information itself as having connectivity. For example, colour information is a connector. Imagine a scattering of dots, some black, some red. The red ones happen to form a larger shape, an X. That X stands out because dots of a similar colour are easily linked together to form a single, integrated representation. Ultimately there is an anatomical underpinning to that phenomenon, but one can make partial sense of it purely from the point of view of information. In the lattice of information, some dots are connected to each other because they are connected to the same colour information. Colour, as a connector, is obviously limited to the visual domain.

Spatial location is another connector, but unlike colour it can operate across sensory domains. If a visual stimulus and a sound appear at the same location, we are prone to link the two, constructing an integrated representation of an object that has both visual and auditory aspects. This spatial interaction has been especially studied in the superior colliculus, where tactile, visual, and auditory information is processed in a single spatial framework (Stein, Stanford and Rowland, 2009). But even though location information can be used to link across sensory domains, it is still limited in its ability to bridge some kinds of information. Information domains that do not have an obvious spatial component are not included.

An attention schema can act as the ultimate connector. Almost all kinds of information in the brain are subject to attention. An attention schema, a central representation of attention, could serve as a hub that connects to any information domain. Awareness, as a model of attention, is like a colour that can tint any topic. In Figure 1D, the attention schema links information about the self, including the body schema and autobiographical knowledge, with information about an apple, including shape and colour and location. In that way, an overarching representation subsumes a great range of information domains. But the figure diagrams only one limited example. One could just as well attend to a thought, a taste, a recalled memory, an emotion, a movement of the limbs, or even all of those together if you are grasping the apple, biting it, thinking about it, and enjoying it. All of those radically different information domains can be linked to the attention schema and thus linked to each other in a single representation. In effect, the brain constructs an integrated representation of its world at that moment, and the attention schema is the crucial bridge that connects the disparate parts because the attention schema represents the deep, computational relationship between the self and the various components of one's world.

So many people have noted the apparent relationship between consciousness and the widespread integration of information that one can't help thinking, with all the smoke, there must be fire. Surely a good theory of consciousness should include that relationship. The attention schema theory does so quite naturally. But in this theory, consciousness is not magically caused by integrating a mass of information together, like in the science fiction trope where Skynet wakes up. Instead, consciousness is a construct. It is an attention schema. That construct serves a central role in bridging across disparate domains of information, allowing for a more complete model of oneself operating in the world.

3.2. Control of attention

A primary function of the attention schema may be the efficient control of attention.

A basic principle of control theory is that a control system benefits from an internal model of the thing to be controlled (Camacho and Bordons Alba, 2004). For example, the brain constructs a body schema, an internal model of the body, to help control movement (Wolpert, Ghahramani and Jordan, 1995). Like all internal models, the body schema is imperfect. It can sometimes become misaligned from the body. Almost all experimental work demonstrating the existence of the body schema relies on those inaccuracies. When misalignment between the body schema and the body occurs, movement of the body is still possible but the controller suffers characteristic deficits that reveal the importance of the body schema (Graziano and Botvinick, 2002; Scheidt *et al.*, 2005; Wolpert, Ghahramani and Jordan, 1995).

If the attention schema theory is correct, the relationship between consciousness and attention should also adhere to the predictions of control theory. The most experimentally revealing conditions should occur when the internal model makes a mistake — when consciousness becomes misaligned from attention. In that case, the control of attention should suffer in a manner consistent with the loss of an accurate internal model.

In the scientific literature on the relationship between consciousness and attention, typically the term 'awareness' is used to refer to a conscious, reportable experience of the sensory stimulus. In keeping with that usage, in the following discussion I will use the term awareness as synonymous with consciousness. It is now well established that attention and awareness can be separated. People can attend to a stimulus in the absence of awareness of that stimulus (Hsieh, Colas and Kanwisher, 2011; Jiang *et al.*, 2006; Kentridge, Nijboer and Heywood, 2008; Koch and Tsuchiya, 2007; Lamme, 2004; McCormick, 1997; Norman, Heywood and Kentridge, 2013; Tsushima, Sasaki and Watanabe, 2006). This separability has led to the suggestion that attention and awareness may be independent processes. In the attention schema theory, the two are not truly independent. They have a principled relationship, diagrammed in Figure 2. Attention has a control system and one part of that system is an internal control model. That internal model, the attention schema, contains information about awareness. If the system is attending to some item X but has not constructed an awareness of X, that corresponds to a temporarily faulty internal model. The internal model of attention has failed to update correctly. In that case, the control of attention should suffer. Attention should still be possible, but it should lose stability and be more easily perturbed by outside influences. In much the same way, if the body schema fails to register the location of your arm perhaps because of anaesthesia of the arm, of course you still have an arm, and you can even control its movement to some degree; but the arm becomes less stable and more easily perturbed by outside influences. Here it is useful to make some clarifications. In control theory, the internal model is not the entire controller. It is one useful part of the controller. Without it, or if it becomes impaired, some control is still possible. Control is compromised in specific ways that are discussed in greater detail below.

Most of the experiments that distinguish visual attention and visual awareness have focused on the first-order phenomenon: attention can exist without awareness. Few experiments ask whether attention behaves the same way, or changes, when awareness is present or absent. The paradigms used to manipulate visual awareness typically involve major changes to the stimulus. As a result, the aware condition and the unaware condition are not easy to compare quantitatively.

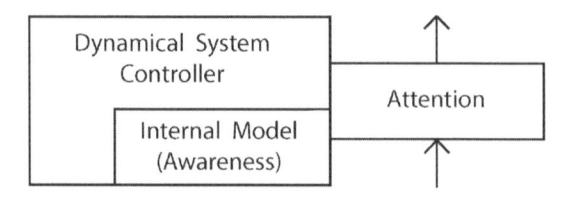

Figure 2. Awareness as an internal control model for attention. The arrows represent information subjected to the process of attention. Attention is regulated by a complex, dynamical systems controller. In the attention schema theory, one part of that controller is an internal model of attention, and awareness is the cartoonish depiction of attention rendered by that internal model.

Recently we performed a series of psychophysics experiments in human subjects to answer this question (Webb, Kean and Graziano, 2016). In those experiments, a visual stimulus drew people's attention to a location. In one condition the stimulus was masked such that participants were subjectively unaware of it. In another condition the mask was adjusted such that participants reported being subjectively aware of the stimulus. The amount of attention drawn by the stimulus was measured using a standard Posner paradigm.

On the basis of the attention schema theory, we predicted that attention would show less stability over time when awareness of the stimulus was absent. This prediction was confirmed. Without awareness of the stimulus, attention to that stimulus behaved in a less stable manner. Attention wobbled up and down significantly more during the tested time interval. Figure 3 shows data from one of the experiments that demonstrates this stabilizing effect of awareness on attention. From the point of view of control theory, when awareness of the stimulus was absent, attention to the stimulus acted as though the stabilizing, internal control model of attention was missing. These experiments are among the most direct tests of the hypothesis that awareness serves as the internal model for attention.

A separate line of experiments by Schurger and colleagues (2010; 2015) also supports the attention schema theory. In those experiments, a visual stimulus evoked activity throughout the visual cortex. Neuronal representation was more stable in time and more consistent across trials when awareness was present than when awareness was absent. Experiments such as these point toward the attention schema theory, in which awareness plays a fundamental role as the internal, stabilizing control model for attention.

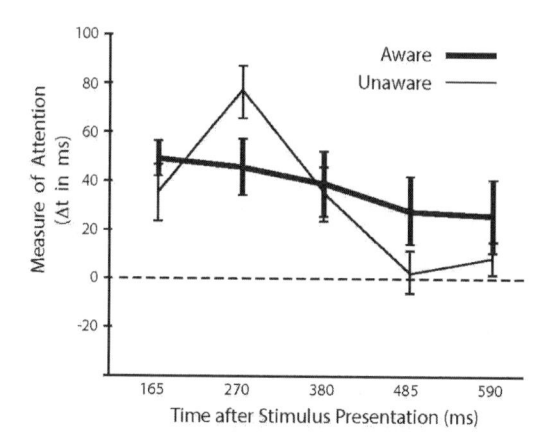

Figure 3. Testing attention with and without awareness. In this experiment, attention to a visual stimulus is tested by using the stimulus as a cue in a Posner spatial attention paradigm (see Webb, Kean and Graziano, 2015, for details). In some trials, the participants are aware of the visual cue (thick line). In other trials, they are unaware of it (thin line). Attention to the cue is less stable across time when awareness is absent. This result follows the predictions of control theory in which an internal control model helps to maintain stability of the controlled variable. The X axis shows time after cue onset. The Y axis shows attention drawn to the cue (Δt = [mean response time for spatially mismatching trials in which the test target appeared on the opposite side as the initial cue] − [mean response time for spatially matching trials in which the test target appeared on the same side as the initial cue]). Error bars are standard error.

One of the central questions of consciousness research is whether consciousness has any adaptive role and, if so, what that role may be. In the attention schema theory, consciousness serves a set of well-defined and testable roles in information processing. Its most basic role, perhaps its evolutionary origin, is to serve as a control model for attention. An engineer who wishes to understand how the brain processes information must understand the internal models that the brain uses to regulate itself. In this theory, consciousness is one of those internal models.

3.3. Social cognition

A third possible adaptive use of an attention schema lies in social cognition. I will begin with an analogy to the body schema. The brain evolved an internal model of the body, an ever-changing, ever-updating complex of information that describes the physical shape, structure, and movement of the body (Graziano and Botvinick, 2002). The body schema presumably first evolved when nervous systems became sophisticated in the control of movement, probably more than half a billion years ago.

One often overlooked function of the body schema, in humans, is to model the bodies of others. If subjects look at a picture of a hand and decide whether it is a left or right hand, the decision is markedly faster when the pictured hand already matches the configuration of the subject's own hand (Parsons, 1987; Sekiyama, 1982). The more different the configurations, the longer the latency to respond, as though the subjects were mentally reconfiguring their own hands to match the picture. Moreover, the same cortical areas, especially the posterior parietal lobe, were active whether judging other people's body configurations or one's own body schema (Bonda et al., 1995). Presumably the use of the body schema to model oneself and control one's own movements emerged first in evolution, and its use in monitoring and predicting the bodies of others was a gradual evolutionary extension.

In the attention schema theory, a similar extension of function occurred for the internal model of attention. In the theory, the attention schema first evolved as a simple model that was part of the control mechanism for attention. Its function was to model one's own state of attention. However, over evolutionary time, an additional adaptive function emerged. The attention schema became increasingly adapted to model the attentional states of others. The advantage is obvious. It improves my ability to predict the behaviour of others and thus guide my own behaviour with respect to others. Attention is one of the main determinants of behaviour. If Bill's brain is focusing attention on X, then X is likely to dominate his behaviour. If I want to predict Bill's behaviour, it would be useful for me to have a model of attention that I can apply to Bill — a model of what attention is, what its dynamics and consequences are, and what in particular Bill is attending to. In the theory, the awareness that I attribute to Bill is a simplified and effective model of his deep attentive processing of information.

In this proposal, we do not merely figure out intellectually whether someone is aware of something. Instead that attribution has an automatic, immediate, perceptual quality because it depends on an internal model constructed beneath the level of high-order cognition.

Ventriloquism is a good example that helps to demonstrate the contrast between a cognitive model and perceptual model of someone else's mind. We have a *perception* of awareness in the ventriloquist puppet, while at the same time we know cognitively that the puppet is not really aware.

People attribute awareness to more than just other people and ventriloquist dummies. We attribute it to our dogs and cats. Some people feel that their houseplants are conscious. Animism is the attribution of subjective consciousness to trees, rivers, volcanoes, and other natural phenomena. Gods, angels, ghosts, spirits are all attributions of awareness. The belief in life after death is the false attribution of awareness outside the physical body. When I am alone in a dark house on a stormy night I sometimes have a creepy feeling that another conscious agent is in the next room stalking me. Intellectually I know it's not true, but my brain has evidently constructed that perceptual model. Sometimes people get angry at the car or the coffee machine as if it were aware of its misdeeds. I argue that this secondary role for the attention schema, attributing awareness to others, is the primary basis for human spiritual belief. We live in a world awash in perceived awareness.

Psychophysical evidence suggests that people do indeed construct a rich model of the attentional state of others. Most research on the topic focuses on only one component, the gaze direction of others. However, people do combine gaze cues, facial expression, and context when assessing the attentional state of others (Kelly *et al.*, 2014). A recent study suggests that we not only reconstruct the object of someone else's attention, we also reconstruct some of the dynamic aspects of attention such as whether attention was drawn extrinsically (by an external stimulus) or directed intrinsically (by an internal decision) (Pesquita, Chapman and Enns, forthcoming). These studies are beginning to show that the brain does indeed construct a rich internal model of the attentional state of others, consistent with the attention schema theory.

One particular brain area may be a central node in a network that computes information related to awareness. Damage to the temporoparietal junction (TPJ) can result in hemispatial neglect, a disturbance of awareness (Valler and Perani, 1986). Attributing states of awareness to others evokes activity in the same subregions of the TPJ (Kelly *et al.*, 2014; Igelstrom *et al.*, 2016). These studies of course do not pin a single function on the TPJ, which presumably contributes to a range of cognitive functions. The studies do, however, begin to support the

4. Illusionism

In the target article of this special issue, Frankish describes an approach to consciousness called illusionism that is shared by many theories of consciousness. The attention schema theory has much in common with illusionism. It clearly belongs to the same category of theory, and is especially close to the approach of Dennett (1991). But I confess that I baulk at the term 'illusionism' because I think it miscommunicates. To call consciousness an illusion risks confusion and unwarranted backlash. To me, consciousness is not an illusion but a useful caricature of something real and mechanistic. My argument here concerns the rhetorical power of the term, not the underlying concepts.

In my own discussions with colleagues, I invariably encounter the confusion and backlash. To most people, an illusion is something that does not exist. Calling consciousness an illusion suggests a theory in which there is nothing present that corresponds to consciousness. However, in the attention schema theory, and in the illusionism described by Frankish, something specific is present. In the attention schema theory, the real item that exists inside us is covert attention — the deep processing of selected information. Attention truly does exist. Our internal model of it lacks details and therefore provides us with a blurred, seemingly magicalist account of it.

Second, in normal English, to experience an illusion is to be fooled. To call consciousness an illusion suggests to most people that the brain has made an error. In the attention schema theory, and also in the illusionism approach described by Frankish, the relevant systems in the brain are not in error. They are well adapted. Internal models always, and strategically, leave out the unnecessary detail.

Third, most people understand illusions to be the result of a subjective experience. The claim that consciousness is an illusion therefore sounds inherently circular. Who is experiencing the illusion? It is difficult to explain to people that the experiencer is not itself conscious, and that what is important is the presence of the information and its impact on the system. The term illusion instantly aligns people's thoughts in the wrong direction.

All of the common objections I encounter have answers. They are based on a misunderstanding of illusionism. But the misunderstanding is my point. Why use a misleading word that requires one to backtrack and explain? For these reasons, in my own writing I have avoided calling consciousness an illusion except in specific circumstances, such as the consciousness we attribute to a ventriloquist puppet, in which the term seems to apply more exactly.

Perhaps I am too much of a visual physiologist at heart. To me, an illusion is a mistake in a sensory internal model. It introduces a consequential discrepancy between the internal model and the real world. That discrepancy can cause errors in behaviour. In contrast, an internal model, at all times, with or without an illusion, is an efficient, useful compression of data. It is never literally accurate. Even when it is operating correctly and guiding behaviour usefully, it is a caricature of reality. I am comfortable calling consciousness a caricature, but not an illusion. It is a cartoonish model of something real.

References

Baars, B.J. (1988) *A Cognitive Theory of Consciousness*, New York: Cambridge University Press.
Beck, D.M. & Kastner, S. (2009) Top-down and bottom-up mechanisms in biasing competition in the human brain, *Vision Research*, **49**, pp. 1154–1165.
Bonda, E., Petrides, M., Frey, S. & Evans, A. (1995) Neural correlates of mental transformations of the body-in-space, *Proceedings of the National Academy of Sciences USA*, **92**, pp. 11180–11184.
Camacho, E.F. & Bordons Alba, C. (2004) *Model Predictive Control*, New York: Springer.
Crick, F. & Koch, C. (1990) Toward a neurobiological theory of consciousness, *Seminars in the Neurosciences*, **2**, pp. 263–275.
Damasio, A.R. (1990) Synchronous activation in multiple cortical regions: A mechanism for recall, *Seminars in the Neurosciences*, **2**, pp. 287–296.
Dennett, D.C. (1991) *Consciousness Explained*, Boston, MA: Little, Brown, & Co.
Desimone, R. & Duncan, J. (1995) Neural mechanisms of selective visual attention, *Annual Review of Neurosciences*, **18**, pp. 193–222.
Engel, A.K. & Singer, W. (2001) Temporal binding and the neural correlates of sensory awareness, *Trends in Cognitive Sciences*, **5**, pp. 16–25.
Frankish, K. (this issue) Illusionism as a theory of consciousness, *Journal of Consciousness Studies*, **23** (11–12).
Gazzaniga, M.S. (1970) *The Bisected Brain*, New York: Appleton Century Crofts.
Graziano, M.S.A. (2013) *Consciousness and the Social Brain*, New York: Oxford University Press.
Graziano, M.S.A. (2014) Speculations on the evolution of awareness, *Journal of Cognitive Neuroscience*, **26**, pp. 1300–1304.
Graziano, M.S.A. & Botvinick, M.M. (2002) How the brain represents the body: Insights from neurophysiology and psychology, in Prinz, J. & Hommel, B. (eds.) *Common Mechanisms in Perception and Action: Attention and Performance XIX*, pp. 136–157, Oxford: Oxford University Press.

Graziano, M.S.A. & Kastner, S. (2011) Human consciousness and its relationship to social neuroscience: A novel hypothesis, *Cognitive Neuroscience*, **2**, pp. 98–113.

Hsieh, P., Colas, J.T. & Kanwisher, N. (2011) Unconscious pop-out: Attentional capture by unseen feature singletons only when top-down attention is available, *Psychological Science*, **22**, pp. 1220–1226.

Igelstrom, K., Webb, T.W., Kelly. Y.T. & Graziano, M.S.A. (2016) Topographical organization of attentional, social and memory processes in the human temporoparietal cortex, *eNEuro*, **3**, ENEURO.0060-16.2016.

Jiang, Y., Costello, P., Fang, F., Huang, M. & He, S. (2006) A gender- and sexual orientation-dependent spatial attentional effect of invisible images, *Proceedings of the National Academy of Sciences USA*, **103**, pp. 17048–17052.

Kelly, Y.T., Webb, T.W., Meier, J.D., Arcaro, J. & Graziano, M.S.A. (2014) Attributing awareness to oneself and to others, *Proceedings of the National Academy of Sciences USA*, **111**, pp. 5012–5017.

Kentridge, R.W., Nijboer, T.C. & Heywood, C.A. (2008) Attended but unseen: Visual attention is not sufficient for visual awareness, *Neuropsychologia*, **46**, pp. 864–869.

Koch, C. & Tsuchiya, N. (2007) Attention and consciousness: Two distinct brain processes, *Trends in Cognitive Sciences*, **11**, pp. 16–22.

Lamme, V.A. (2004) Separate neural definitions of visual consciousness and visual attention: A case for phenomenal awareness, *Neural Networks*, **17**, pp. 861–872.

Lamme, V.A. (2006) Towards a true neural stance on consciousness, *Trends in Cognitive Sciences*, **10**, pp. 494–501.

Lau, H. & Rosenthal, D. (2011) Empirical support for higher-order theories of consciousness, *Trends in Cognitive Sciences*, **15**, pp. 365–373.

McCormick, P.A. (1997) Orienting attention without awareness, *Journal of Experimental Psychology, Human Perception and Performance*, **23**, pp. 168–180.

Metzinger, T. (2010) *The Ego Tunnel*, New York: Basic Books.

Newman, J. & Baars, B.J. (1993) A neural attentional model for access to consciousness: A global workspace perspective, *Concepts in Neuroscience*, **4**, pp. 255–290.

Nisbett, R.E. & Wilson, T.D. (1977) Telling more than we can know — verbal reports on mental processes, *Psychological Review*, **84**, pp. 231–259.

Norman, L.J., Heywood, C.A. & Kentridge, R.W. (2013) Object-based attention without awareness, *Psychological Science*, **24**, pp. 836–843.

Parsons, L.M. (1987) Imagined spatial transformations of one's hands and feet, *Cognitive Psychology*, **19**, pp. 178–241.

Pesquita, A., Chapman, C.S. & Enns, J.T. (forthcoming) Seeing attention in action: Human sensitivity to attention control in others, *Proceedings of the National Academy of Sciences USA*.

Scheidt, R.A., Conditt, M.A., Secco, E.L. & Mussa-Ivaldi, F.A. (2005) Interaction of visual and proprioceptive feedback during adaptation of human reaching movements, *Journal of Neurophysiology*, **93**, pp. 3200–3213.

Schurger, A., Pereira, F., Treisman, A. & Cohen, J.D. (2010) Reproducibility distinguishes conscious from nonconscious neural representations, *Science*, **327**, pp. 97–99.

Schurger, A., Sarigiannidis, I., Naccache, L., Sitt, J.D. & Dehaene, S. (2015) Cortical activity is more stable when sensory stimuli are consciously perceived, *Proceedings of the National Academy of Sciences USA*, **112**, pp. E2083–2092.

Sekiyama, K. (1982) Kinesthetic aspects of mental representations in the identification of left and right hands, *Perceptual Psychophysics*, **32**, pp. 89–95.

Stein, B.E., Stanford, T.R. & Rowland, B.A. (2009) The neural basis of multisensory integration in the midbrain: Its organization and maturation, *Hearing Research*, **258**, pp. 4–15.

Szczepanowski, R. & Pessoa, L. (2007) Fear perception: Can objective and subjective awareness measures be dissociated?, *Journal of Vision*, **10**, pp. 1–17.

Tononi, G. (2008) Consciousness as integrated information: A provisional manifesto, *Biological Bulletin*, **215**, pp. 216–242.

Tsushima, Y., Sasaki, Y. & Watanabe, T. (2006) Greater disruption due to failure of inhibitory control on an ambiguous distractor, *Science*, **314**, pp. 1786–1788.

Vallar, G. & Perani, D. (1986) The anatomy of unilateral neglect after right-hemisphere stroke lesions: A clinical/CT-scan correlation study in man, *Neuropsychologia*, **24**, pp. 609–622.

Webb, T.W. & Graziano, M.S.A. (2015) The attention schema theory: A mechanistic account of subjective awareness, *Frontiers in Psychology*, doi: 10.3389/fpsyg.2015.00500.

Webb, T.W., Kean, H.H. & Graziano, M.S.A. (2016) Effects of awareness on the control of attention, *Journal of Cognitive Neuroscience*, **28**, pp. 842–851.

Wolpert, D.M., Ghahramani, Z. & Jordan, M.I. (1995) An internal model for sensorimotor integration, *Science*, **269**, pp. 1880–1882.

Nicholas Humphrey

Redder than Red

Illusionism or Phenomenal Surrealism?

Abstract: *Sensations represent our subjective 'take' on sensory stimulation — how we feel about red light falling on the retina, salt dissolving on the tongue, a thorn piercing the skin. They tell — in the language of phenomenal properties — what the experience is like for us. In so far as they represent the reality of this subjective relationship, they cannot be said to be illusory. The relationship, magical as it may seem, is not being misrepresented as something it is not. If anything, it is being represented as something 'super-real'.*

An early draft of Keith Frankish's important paper was called 'The Magic Problem'. Later it became 'The Illusion Problem', before settling on the title it has now. I have to say I wish he had stuck with magic. It's true that illusion and magic have overlapping meanings. But illusion is generally defined by what's *wrong* with it. According to the *Oxford English Dictionary* online[1] an illusion is: 'An instance of a wrong or misinterpreted perception.' 'A deceptive appearance or impression', 'A false idea or belief'. Magic by contrast is defined by what's *right* with it. Magic is: 'a power that is apparently mysterious or supernatural', 'a trick performed for entertainment', 'a quality of being beautiful and delightful in a way that seems remote from daily life'. As a descriptor of consciousness, magic fits the bill much better than illusion.

I acknowledge this is revisionary on my part. As Frankish notes, I have in the past explicitly endorsed illusionism. I've argued that con-

Correspondence:
Nicholas Humphrey, Darwin College, Cambridge. *Email:* humphrey@me.com

---

[1] http://www.oxforddictionaries.com/

sciousness involves a creative misrepresentation of sensory stimulation at the body surface. And I have used the analogy of the 'real impossible triangle' to illustrate how this could work: how we could be deceived by a specially constructed brain state into believing we are in the presence of phenomena that do not actually exist. But — partly spurred by Frankish's article — I've now come to think that this emphasis on the illusory content of sensations is unhelpful, and even conceptually muddled. It rests on a failure — by me as much as Frankish — to appreciate just what sensations are about and so what exactly it is that seems to have phenomenal qualities.

This is remiss of me, because the theoretical ground was well prepared. I've long maintained that sensations are representations of something *we do* (Humphrey, 1992; 2006; 2011). The story begins far back in time with creatures — our distant ancestors — making reflex motor responses to stimuli arriving at their bodies: *expressive responses*, 'wriggles of acceptance or rejection' as I've called them. These responses would have been tailored by natural selection to be biologically adaptive — taking account of the nature of the stimuli, where on the body they were arriving, and what importance they had for the creature's well-being. Dennett has stipulated that 'No afferent can be said to have a significance "A" until it is "taken" to have the significance "A" by the efferent side of the brain' (Dennett, 1969, p. 74). But, from early on, this condition would have been satisfied: the responses were a 'take' on the meaning of the stimulation, what it was *about*. To start with, however, there would have been no one at home in the brain, no subject, capable of accessing this 'aboutness'. The next step in the evolution of sensations would be for selection to find a way for the creature to extract the meaning from its own expressive behaviour. And, as it turned out, there was a neat solution available: this was to develop a special brain module whose job was to *read* the efferent activity at the level of motor command signals being sent downstream.

Fast forward to where we are today. The upshot is that we human beings are still discovering meaning in sensory stimulation in much the same way. When our specialized sense organs detect potentially significant stimuli — light falling on our retinae, sound reaching our eardrums, heat burning our skin, and so on — our brains still reflexly formulate evaluative responses. And it is by monitoring how our brains are responding that we come to have conscious sensations of light, sound, pain. True, the responses are no longer quite what they once were. In the course of history they have become internalized —

or as I like to say, 'privatized' — so that they no longer result in overt bodily behaviour. They are now as-if responses expressed by a virtual body. But they are still there. And the reason they are still there is precisely that they underwrite the reading that translates into conscious experience.

So, what bearing does this evolutionary history have on illusionism and Frankish's paper? I'd say the lesson is that, when considering whether sensations are or are not 'real', we must never let go of the fact that sensations do indeed represent *our take* on stimuli impinging on the body. In doing so they represent some of the objective facts about what's happening: the what, where, and when, for example. But, crucially, they also represent how we *evaluate* what's happening, how we *feel* about it. And this is where phenomenal properties come into their own. Sensations represent how we relate to stimulation using, as it were, a paintbox of phenomenal concepts to depict what it's like for us: what it's like to have red light falling on the retina, salt dissolving on the tongue, a thorn piercing the skin, and so on.

Consider a prototypical experience of your own. Suppose a beetle is crawling across the skin of your back. Your brain reacts to signals from your skin with an ancient internalized response, which, when you read it, yields a sensation of being touched. On the objective side, the sensation represents the touch event as having certain spatial and temporal properties that tell of where on your skin it's occurring, how long it lasts, the spatial pattern, and so on. Meanwhile, on the subjective side, it represents the event as having a range of phenomenal properties that tell of *what it's like*: it is yours (and no one else's), it has a distinctly tactile feel (as opposed to visual, say), it's ticklish, you feel like slapping it... and, yes, it's 'simple, ineffable, intrinsic, private, immediately apprehended' (not to mention 'inaccessible to third-person science, and inexplicable in physical terms' — Frankish, this issue, p. 13).

Then where could *illusion* come into this? What grounds could there possibly be for suggesting that the represented properties are actually not what they appear to be? *It depends on which aspect of the sensation we're talking about.*

It's certainly possible you might be getting a false idea about the objective facts. There is, for example, a tactile illusion called the 'Cutaneous Rabbit': 'A rapid sequence of taps delivered first near the wrist and then near the elbow creates the sensation of sequential taps hopping up the arm from the wrist towards the elbow, although no physical stimulus was applied between the two actual stimulus locations' (https://en.wikipedia.org/wiki/Cutaneous_rabbit_illusion). So it is possible (even if unlikely in the case of the beetle) you might be misled about the spatial location of the stimuli.

However, in what circumstance, if any, could you be getting a false impression about what it feels like? How could you — as Frankish seems to suggest — be experiencing a feel that 'doesn't exist'? To be blunt, I think the very notion of this is absurd. When the sensation represents you as feeling a certain way about the stimulation, *that is all there is to it.* The phenomenal feel arises with the representation, and *thereby its existence becomes a fact.* When the sensation represents the feel as ticklish, so it does. When it represents it as intrinsic, ineffable, etc., so it does. When it represents it as being like something magical that no one can explain, again this is just what it does. Phenomenal properties leave behind no residue that could separately be assessed as a misrepresentation of the reality. Dennett has remarked that phenomenal qualia are like 'a beautiful discussion of purple, just about a colour, without itself being coloured' (Dennett, 1991, p. 371). But, discussions — and the opinions expressed in them — are what they are. There's no way an opinion can be illusory.

I was wrong, then, in my earlier writing, to draw an analogy between phenomenal consciousness and the real impossible triangle, which is a clear case where we make a mistake about the facts. I ought instead to have chosen a very different kind of 'magical' entity: one where the magic lies not in what we perceive it to be as an independent object but in how we are affected by it subjectively. A flower, for example, that we find *beautiful*. An action that we find morally *good*. Or, to lower the bar, a joke that we find *funny*. None of these higher-order relational attributions are *illusory*. Take music. 'Is it not strange that sheep's guts could hale souls out of men's bodies?' (Shakespeare, 1623, 2.3.62). Yes, it is, wonderfully strange. But not a mistake.

Does this mean I now want to say that phenomenal properties really are *real*? Frankish says that when he talks of phenomenal properties as 'not being real or not existing', he means 'that they are not *instantiated* in our world' (p. 12). It's far from clear what he thinks would or would not count as being 'instantiated in our world', especially when what we're talking about is the property of a relationship. Maybe he'd take the line that only the properties of *things* can be real in this sense, and there's no such *thing* as a relationship. But this would be setting the bar for realism implausibly high. It would mean denying that a relationship such as a marriage, for example, can have real properties. Or, more to the point, it would mean denying that any of our own subjective attitudes to events, such as finding a joke funny, can do. And I don't think Frankish would want to go that way. Rather, I expect he would agree that a property such as 'being funny', which is essential to our take on the joke, must actually be an inherent feature of the brain activity that brings the representation about. This brain activity is the cause of everything at the level of language and behaviour that constitutes our finding the joke funny. That's not to say that 'being funny' is a property *of* the brain activity. But a third-person observer could in principle recognize what the physical brain activity means to the subject — and surely that makes funniness real enough.

Then I expect — at any rate I hope — Frankish would also agree that the phenomenal properties of sensations must likewise be an inherent feature of the brain activity that brings the representation about. I'd go further. If sensation is in fact a reading of our own internalized motor response to stimulation, then we might expect that phenomenal properties will be correlated with the *adverbial quality* of this response (Humphrey, 1992). We are responding redly, saltly, painily... magically. Now, when we monitor these responses from the inside, we are representing the adverbial quality as the *modality* of the phenomenal feel we attribute to the stimulation. One day in the future, when scientists monitor the same activity by means of a brain scan — and monitor us monitoring it — I'd predict these phenomenal properties will indeed show up 'for real' from the outside.

So, yes, I have become a realist of sorts about phenomenal properties. However, I hasten to say this does not make me a realist of the kind Frankish so skilfully skewers in his paper. In fact, I have a particular reason for wanting to distance myself from those other realists, and this is that they, just like illusionists, have the wrong target in their sights. According to a flyer for the Tucson Consciousness Conference 2016, realists — a.k.a. 'philosophical dualists, panpsychists, spiritualists, and proponents of quantum brain biology' — suggest that 'mental qualities or conscious precursors are somehow intrinsic features of the universe, that consciousness has, in some sense, been here all along' (Tucson, 2016). If this means anything at all, I assume it means that they think of consciousness as having an absolute objective reality. They clearly don't think of it as emerging in one small corner of the universe as a property of the *take* that evolved creatures have on stimulation of their bodies. Nagel, Chalmers,

Strawson, and others in the realist camp are realists about the wrong thing.

To sum up. Neither illusionism nor realism addresses what should be the central question for a theory of consciousness: namely, how we represent the meaningful relationship we have to sensory stimulation. What, then, should we call a theory that attempts to explain how the properties of this relationship come to be real, true, and magically deep? I have a suggestion: *phenomenal surrealism* — where 'surreal' has the meaning that Picasso originally gave it. 'What I intended when I invented this word, [was] something more real than reality... Resemblance is what I am after — a resemblance deeper and more real than the real, that is what constitutes the sur-real' (Picasso, 1933/2007). It was in this spirit that Picasso could say of his great sculpture of a goat, 'She's more like a goat than a real goat, don't you think'.

My thought, then, is this: just as Picasso's goat was goatier than a real goat, so phenomenal redness is redder than real red, phenomenal pain painier than real pain. In general phenomenal properties are represented in sensation as 'more real' than the actual physiological events that give rise to them. By adding in the relational dimension of how we feel about it, sensation has, as it were, put one over on the physical reality of stimulation. To invoke the definition of 'magic' I began with, what is being added is 'a quality of being beautiful and delightful in a way that seems remote from daily life': except of course that, for we creatures who are fortunate enough to have evolved phenomenal consciousness, *it is daily life*. I like to think the artist Samuel Palmer was making the same philosophical point when he wrote in his Journal, 'Bits of nature are generally much improved by being received into the soul' (Palmer, 1824). Palmer's 1830 painting of the 'Magic Apple Tree' depicts a tree bending under the weight of fruit, and still more under *the weight of the fruit's colour*.

Is this just about words? Does it matter whether we call our theory of consciousness illusionism, or surrealism? I wish I could say no. But I'm not so sanguine. I worry that illusionism — unbridled illusionism such as Frankish champions in his article — threatens to undermine the very cause to which we are all committed: namely, to provide a scientific materialist explanation of consciousness which is both true and persuasive.

I worry that illusionism, as a theoretical framework, gets the science of consciousness off on the wrong foot. If taken literally, it points us towards trying to explain the phenomenal properties of consciousness on the model of explaining *why a funny joke isn't actually funny*. And this, as we saw above, is not going anywhere. By contrast phenomenal surrealism, if we were to adopt it, would send us in a much more promising direction. It would suggest we try to explain phenomenal properties on the model of explaining *why a joke becomes funny for us*. I've taken this second approach in 'A Riddle Written on the Brain' (Humphrey, 2016).

And then I worry that, when it comes to winning the argument in the forum of public opinion, in-your-face illusionism is bad politics. Illusionism cannot but feed ordinary people's fears that we scientist/ philosophers *want to take consciousness away from them*. How else should we expect people to react when we tell them — the words are Frankish's — 'if we mean [conscious] experiences with phenomenal properties, then illusionists do indeed deny that such things exist' (p. 21) . Whatever Frankish might want readers to understand by this — and of course his discussion is actually thoughtful and nuanced — it is an invitation to many people to stop listening. Anti-materialist philosopher Mary Midgley, for example, will write a book titled 'Am I an Illusion?', and think she can dispose of our arguments by pinching herself and saying in effect 'don't be daft'.

With phenomenal surrealism, on the other hand, the message would be just the opposite. Rather than taking something away from people, we would be adding to their estimate of who and what they are. We'd be reminding them that consciousness is their own creation. And, more than that, we'd be encouraging them to think of themselves as artists — creators of experiences that are 'more real than real'. In response to Midgley, we could say: 'Don't worry. You are an astonishing work of art' (Humphrey, 2015).[2]

---

[2] But is there not some part of our conscious experience that is genuinely illusory? In Section 1.4 of his article Frankish briefly discusses 'outward-looking illusionism' which he takes to be the mistake of *projecting* the phenomenal properties of sensation onto things in the external world. As he says, 'we both misrepresent features of experience as phenomenal and then re-represent these illusory properties as properties of the external world, mistaking complex physical properties of our sensory states for simple phenomenal properties of external objects'. For the reasons given, I cannot agree that the first level of representing is a misrepresentation. But the second level of re-representing is a very different matter. If and when we project subjective phenomenal properties onto external objects, this truly is a misrepresentation. We are mistaking facts about how we ourselves feel about sensory stimulation for facts about the objects that give rise to the stimulation. The result is we may indeed come under the illusion — and this really is an illusion — that things out there in the world are 'singing our song' (Humphrey, 2011, chapter 7).

References

Dennett, D.C. (1969) *Content and Consciousness*, London: Routledge & Kegan Paul.

Dennett, D.C. (1991) *Consciousness Explained*, Boston, MA: Little Brown.

Humphrey, N. (1992) *A History of the Mind*, London: Chatto & Windus.

Humphrey, N. (2006) *Seeing Red: A Study in Consciousness*, Cambridge, MA: Harvard University Press.

Humphrey, N. (2011) *Soul Dust: The Magic of Consciousness*, Princeton, NJ: Princeton University Press.

Humphrey, N. (2015) Consciousness as art, *Scientific American Mind*, May/June, pp. 65–69.

Humphrey, N. (2016) A riddle written on the brain, *Journal of Consciousness Studies*, **23** (7–8), pp. 278–287.

Palmer. S. (1824) *Shoreham Notebooks*, quoted in Palmer, A.H. (1892) *The Life and Letters of Samuel Palmer, Painter and Etcher*, pp. 16–17, London: Seeley.

Picasso, P. (1933/2007) quoted in Richardson, J., *A Life of Picasso: The Triumphant Years: 1917–1932*, p. 349, London: Cape.

Shakespeare, W. (1623) *Much Ado About Nothing*, 2.3.62.

Tucson (2016) Online poster, *The Science of Consciousness conference*, 25–30 April 2016.

François Kammerer

The Hardest Aspect of the Illusion Problem — and How to Solve it

Abstract: *In 'Illusionism as a Theory of Consciousness', Frankish argues for illusionism: the thesis that phenomenal consciousness does not exist, but merely seems to exist. Illusionism, he says, 'replaces the hard problem with the illusion problem — the problem of explaining how the illusion of phenomenality arises and why it is so powerful'. The illusion of phenomenality is indeed quite powerful. In fact, it is much more powerful than any other illusion, in the sense that we face a very special and unique intuitive resistance when trying to accept that phenomenality is an illusion. This is bad news for illusionists, because this means that they cannot entirely model their explanation of the illusion of consciousness on the explanation of other illusions. Explaining this unique intuitive resistance to illusionism may therefore constitute the hardest aspect of the illusion problem. However, I think that this aspect of the problem is solvable. I will outline a possible solution, which is based on the hypothesis that our (illusory) introspective representations of phenomenal states characterize them as having unique epistemological properties and as playing a special epistemological role.*

In 'Illusionism as a Theory of Consciousness', Keith Frankish argues that phenomenal consciousness does not exist, but merely seems to exist. I will not try to criticize his arguments here, as I am quite convinced that some kind of illusionism must be true regarding phenomenal consciousness. What I want to do is to examine what I take to be the most difficult challenge for illusionist theories of consciousness,

Correspondence:
François Kammerer, Université Paris-Sorbonne, Paris, France.
Email: kammerer.francois@gmail.com

which does not receive any particular attention in Frankish's paper. I will explain why I think current illusionist theories of consciousness fail to meet this challenge, and I will outline a possible solution to this problem.

1. The hardest aspect of the illusion problem

Proponents of illusionism, says Keith Frankish, replace the hard problem of consciousness with the illusion problem (Frankish, this issue, p. 37). Instead of having to explain how and why phenomenal consciousness arises from the physical, they only have to explain why it *seems to us* that phenomenal consciousness exists, while it does not. They also have to explain why this illusion is so powerful.

While this problem may be much more tractable than the original hard problem of consciousness, I think that this problem has a 'hard' aspect which does not receive a lot of attention in Frankish's paper, and more generally in the work of proponents of illusionism. This aspect consists in the explanation of the following fact: the illusion of phenomenality is not only *very powerful*; it is, in fact, *much more* powerful than *any other illusion*. It is powerful in a very distinctive way, as we face a unique intuitive resistance when we try to accept the illusory nature of phenomenality.

Indeed, denying the existence of phenomenal states does seem crazy to many. Chalmers writes that in doing so one 'denies the evidence of our own experience. This is the sort of thing that can only be done by a philosopher' (Chalmers, 1996, p. 188). Searle writes about Dennett's illusionism that it 'denies the existence of the data'; and he comes close to calling this kind of position insane — 'surely no sane person could deny the existence of feelings' (Searle, 1997). Other examples of such statements abound.[1] And anyone can easily feel the strangeness of illusionism simply by focusing on her own experience. Let's say that you are having an experience of a red rose. Focus on your experience, and then try to think: 'I am not really having an experience endowed with the qualitative feel it seems to have. This is simply an illusion.' Chances are that you will find this idea deeply puzzling

---

[1] Bryan Frances writes, in a discussion of scepticism: 'I assume that eliminativism about feelings really is crazy' (Frances, 2008, p. 241). Galen Strawson writes that eliminativists who deny that there are mental states such that there is something it is like to be in these states 'do seem to be out of their minds', and that their position is 'crazy, in a distinctively philosophical way' (Strawson, 1994, p. 101).

and even hard to make sense of, even if you embrace illusionism from a theoretical point of view.

I think that this very strong intuitive resistance to illusionism has to be explained by illusionist theories. After all, the fact that we are so strongly reluctant to give up our belief in phenomenality characterizes the peculiar way in which we are subjected to the illusion of phenomenality. This intuitive resistance to illusionism is therefore a part of the *explanandum* of illusionist theories. Moreover, this strong intuitive resistance to illusionism, and the fact that it seems crazy, is probably the main reason why illusionism gained so little support amongst philosophers of mind, in spite of its numerous theoretical advantages. Explaining this strong intuitive resistance within an illusionist framework would therefore be useful for illusionists, as it would give them some dialectical leverage when faced with accusations of craziness. 'It is true that our theory seems crazy, but it explains very simply and naturally many things, including the very fact that it seems crazy.'[2] The problem is that, as I will try to show, explaining this strong intuitive resistance is not easy, but quite hard. It is, in fact, the hardest aspect of the illusion problem.

2. Why is this aspect the hardest aspect of the illusion problem?

Many proponents of illusionism, I think, would begin by saying that illusionist theories can *already* explain this intuitive resistance to illusionism in a rather natural way. They would draw a parallel with perceptual illusions. The illusion of phenomenality may arise because of some hardwired features of our introspective device, in such a way that this illusion has some degree of *cognitive impenetrability* (Frankish, this issue, p. 18). It persists even in the face of opposite beliefs. That could explain why, even if we are convinced illusionists, our intuition still pulls us away from illusionism. Similarly, in the Müller-Lyer illusion, we are still intuitively tempted to judge that the two lines have different lengths, even when we *believe* that they do not.

---

[2] This is not to say that everyone would be fully satisfied by such an answer (Chalmers, 1996, pp. 188–9). However, I think that it is undeniable that a good explanation of the strong intuitive resistance to illusion would at least give *some* dialectical force to the illusionist position.

However, I think that the cases are very different on closer examination. The intuitive resistance we face when we try to think that consciousness is an illusion is special, *sui generis*. It cannot simply be explained by the fact that we have an ongoing, perceptual-like, cognitively impenetrable disposition to believe that we are conscious. Therefore, it cannot be explained simply by following the model of the explanation of perceptual illusions.

The difference amounts to this: what makes us reluctant to accept illusionism is not only that we are disposed to believe that we are conscious, it is also that we have difficulties *making sense of the hypothesis that we are not conscious while it seems to us that we are*. We struggle to simply *picture* what it would mean for consciousness to be an *illusion*. When I focus on my current experience of a red rose, I find the idea that I do not really have this precise feeling, but only an *illusion* of this feeling, not only *unlikely to be true*, but deeply weird, puzzling, and almost senseless.

Nothing similar happens in the case of perceptual illusions. Let's imagine, for example, that I am falling prey to the Müller-Lyer illusion, so that I believe that the two lines are as they seem: that they have different lengths (for example, because it is my first time looking at these lines and I have not measured them yet). Even in this situation, I would still easily *understand* what it would mean for me to be victim of an illusion in that case. The idea that the two lines are *not* as they seem, and that they have in fact the same length *in spite of appearances* would make perfect sense to me.

In order to make the contrast even starker, let's take the example of a perceptual belief I *do* hold. I *do* believe that I have two hands right now, on the basis of my perception of my two hands. My perceptual disposition to believe that I have two hands is very strong, and so is my corresponding belief. However, I have absolutely no problem *entertaining* and *understanding* the hypothesis that I may be a one-armed person victim of an hallucination. It would be *very hard* to convince me that it is the case, but the hypothesis, however far-fetched, is not at all difficult to *understand* or to *make sense of*.

This contrast shows that the intuitive resistance to illusionism regarding consciousness is pretty unique, and cannot simply be explained by the persistence of a very strong perceptual-like, cognitively impenetrable disposition to believe that we are conscious. The illusion of phenomenality is particularly powerful; it is much more powerful than other 'classical' perceptual illusions, and it is powerful in a very distinctive way.

This is a problem for illusionist theories of consciousness. Indeed, most recent illusionist theories of consciousness tried to model their explanation of the illusion of phenomenality on the explanation of other illusions. But if the illusion of phenomenality has a distinctive and special force when compared to other illusions, this means that this kind of explanation is doomed to fail. Indeed, this kind of explanation cannot account for the following contrast: we can easily represent to ourselves that a 'non-phenomenal' situation, which we grasp through perception for example, is illusory, while we struggle deeply to understand how a *phenomenal* situation, grasped through introspection, could be illusory.[3] That is why this aspect of the illusion problem — the explanation of the unique intuitive resistance we face when we try to accept illusionism — is the hardest aspect of the illusion problem.

3. Current illusionist theories of consciousness cannot solve the hardest aspect of the illusion problem: case studies

Nicholas Humphrey (Frankish, this issue, p. 17; Humphrey, 2011, chapter 2) states that the illusion of phenomenality is comparable to the visual illusion of impossible objects. Real physical objects, such as the 'Gregundrum' (Frankish, this issue, p. 17), when seen under a certain perspective, can give the illusion that we are facing an *impossible* object: a solid Penrose triangle. The same way, the introspective perspective that we take on some of our neural states is what gives us the illusion that we have *phenomenal sensations*, which are impossible objects (intrinsic, private entities).

The problem, from my point of view, is that we cannot explain the hardest aspect of the illusion problem simply by following this model. Indeed, when facing a Gregundrum, we do not have any particular intuitive resistance to entertaining the hypothesis that the apparent presence of a solid Penrose triangle could merely be an *illusion*. On the contrary, this hypothesis seems perfectly intelligible and sensible — actually, this may be the most natural hypothesis that comes to mind when facing what seems to be an *impossible* object, such as a

---

[3] Here I simply want to emphasize the existence of such a contrast. I will give my own explanation of why this contrast exists in Section 5.

THE HARDEST ASPECT OF THE ILLUSION PROBLEM 129

Penrose triangle. So, this model cannot explain the peculiarity of the illusion of phenomenality.

Similarly, Michael Graziano has an illusionist theory of consciousness, the 'attention schema theory', that states that the brain monitors its own internal attentional processes in a schematic and fallacious way. Instead of representing what is really there — a set of complex attentional processes — it represents itself as being in states of awareness, where awareness is depicted as a 'fluidic substance', an 'experience', a 'sentience' (Graziano, 2013, p. 80); hence the fallacious impression that we are in qualitative and primitive *phenomenal states*. And, because this impression arises from some features of the very architecture of the brain, any theory that goes against this impression is judged very counter-intuitive. Quite in the same way, Graziano explains, Newton's theory of light, according to which white light is not pure and simple, but composed of all the other coloured lights, struck everyone as highly counter-intuitive when first formulated, because the human visual system is hardwired to depict white light as being simple, pure, and primitive.

However, again, such a model cannot account for the unique intuitive resistance we face in the case of illusionism regarding consciousness. Indeed, in the case of Newton, even though people were strongly disposed to believe that white light was pure, simple, and not composed, I do not think they had any difficulty *understanding* what it would mean for white light to be *profoundly different* from what it seemed to be. However surprising and weird this theory must have seemed to them, I do not think they had trouble picturing what Newton's theory meant: that white light *is not at all like our perception of it depicts it*. This thesis may be counter-intuitive, but it is not difficult to make sense of. But, on the other hand, we *do* have trouble fully grasping what it means for consciousness to be an illusion; we *do* struggle to make sense of this idea. Therefore, Graziano's account cannot solve the hardest aspect of the illusion problem.

Derk Pereboom's illusionist theory of consciousness (Frankish, this issue, pp. 17–8; Pereboom, 2011) is quite similar to the theories I just discussed. He suggests that introspection misrepresents phenomenal properties by depicting them as having a qualitative nature that they do not have, quite in the same way sensation represents external objects as having primitive colour properties ('edenic colours' — Chalmers, 2006; Pereboom, 2011) that they do not have. This hypothesis alone is not able to solve the hardest aspect of the illusion problem, for the same reasons I already mentioned. Indeed, if this hypothesis were true, why would we have such difficulty making sense of illusionism regarding consciousness, while illusionism regarding primitive colours seems rather unproblematic? Indeed, the thesis according to which external objects only *appear* to bear primitive colours, but do not really bear them, is quite widely accepted since Galileo and Descartes (and has been on the philosophical market at least since Epicurus). And this thesis does not seem to be particularly difficult to *understand* or to make sense of.

Pereboom then makes an additional hypothesis: maybe illusionism about consciousness seems much less credible than about secondary properties such as colours because 'we cannot check the accuracy of introspection, as we can that of perception, by adopting different vantage points, using measurement instruments, and so forth' (Frankish, this issue, p. 18; Pereboom, 2011, p. 23). However, this hypothesis does not help much. After all, a physicist could build a detector D, to detect a new particle Z, and for a little while she may very well own only *one* copy of this detector, and have no other way to detect Z. The detections of Z made with D could therefore not be checked for accuracy using any other means. Would that mean that our physicist would have huge trouble *making sense* of the hypothesis that D is in fact inaccurate? Would she find such an hypothesis extremely puzzling, and barely intelligible? This seems quite unlikely. But this shows that the inability to check the accuracy of a given source of information cannot explain why subjects find the idea that the information provided by the source is illusory so puzzling and difficult to accept.

4. The theoretical introspection hypothesis

Current illusionist theories of consciousness cannot solve the hardest aspect of the illusion problem, mostly because they try to model their explanation of the illusion of phenomenality on the explanation of other 'classical', perceptual illusions. But the theories resulting from such a modelling cannot do the job. Indeed, the illusion of phenomenality has a distinctive and unique force *when compared with other illusions* — and this cries out for explanation.

I now want to outline how we could solve this difficulty. First, we have to acknowledge the particular force of the illusion of phenomenality, and the unique predicament we find ourselves in when we try to accept that consciousness is an illusion. By acknowledging this difficulty, we do not weaken the illusionist position — quite the contrary, in my opinion.

Second, we have to develop an hypothesis that could account for the unique intuitive resistance to illusionism about consciousness; for example, an hypothesis concerning introspective representations and introspective processes, which could explain why we are so resistant to illusionism. I will now present an hypothesis I think could do the job. Call it the 'Theoretical Introspection Hypothesis' (TIH).[4]

Theoretical introspection hypothesis: suppose that, as proponents of the 'theory-theory of self-awareness'[5] say, introspection (including in this case phenomenal introspection) is a theoretically informed activity, where the relevant theories are our naïve theory of mind and our naïve epistemology. Suppose that these theories are innate and modular, that is to say, informationally encapsulated (though they are informationally related to each other — they may for example be implemented in the same module). Introspective representations of phenomenal states are deeply theoretically informed by these theories.[6]

The TIH states that the content of these theories includes, so to speak, the following statements:

(1) Minds can take up information about states of affairs, and then can use that information to form beliefs about them: states of affairs *appear* to minds.
(2) The way minds do that is that they are *affected* in a certain way, they have certain *experiences*.
(3) The properties of experiences determine what *appears* to the mind, and a state of affairs appears to the mind in virtue of these properties of experiences. For example: an experience of a red circle is an affection of the mind in virtue of which the presence of a red circle appears to the mind.
(4) Take all the cases in which a certain state of affairs A appears *veridically* to a subject S, and consider what all these cases have in common regarding the way in which S is affected. What they have in common is a state E, which is an *experience of A*.

---

[4] For a first suggestion of an hypothesis of this kind, in a different vocabulary and within a different framework, see Kammerer (2016).
[5] For a definition and a critical overview of the theory-theory of self-awareness, see Nichols and Stich (2003, pp. 164–9).
[6] Frankish quickly mentions this kind of view (Frankish, this issue, p. 36).

Something is a part of E *if and only if* this thing is part of the way in which S is affected in *all the cases* in which A appears veridically to a subject S.[7]

(5) Appearances can be fallacious, and a mind can be deceived by the way states of affairs appear. And here is what happens in cases of fallacious appearances: when a subject S has a fallacious appearance of A, S is affected *in exactly the same way* as in cases of veridical appearances of A, except that A is not the case. That is to say, when a subject S has a fallacious appearance of A, it is in state E (E being, and being nothing but, what is common to the way S is affected in all the cases in which A appears veridically to her), but A is not the case.[8]

Of course, our naïve theory of mind/epistemology does not literally *contain* these statements. Naïve theories, when we conceive of them

---

[7] One could then object that according to this thesis there would be states of affairs such that there can be no experiences of them according to our theory of mind, because there is no common element to the way in which subjects are affected when these states of affairs veridically appear to them, while we still commonly and naturally speak of *experiences* of these states of affairs. Let's take, for example, *the presence of a car*. The presence of a car can appear to me through very different affections (for example, what we call 'visual experiences', 'auditory experiences', 'olfactory experiences'), and there is probably nothing common, from an experiential point of view, to all these affections. However, we can still speak of an 'experience of a car'. That seems to contradict the idea that our naïve theory of mind does not allow for such a thing. My answer is the following: 'being an experience of a car' is indeed not an experiential kind *stricto sensu*, but is simply a secondarily constructed kind, which denotes a disjunction of very numerous (in fact, potentially infinite) and distinct 'real' experiential kinds, such as: 'being a visual experience of a white car-shaped thing of that size, in this perspective, at such distance', 'being an olfactory experience of oil, gasoline, with a soupçon of hot metal', etc. Certainly, for each of these 'real' experiential kinds, there is something that is common to the way in which the subjects are affected in all the situations in which the relevant states of affairs veridically appear to them, and in virtue of which these states of affairs appear to them.

[8] So, according to our naïve theory of mind/epistemology, minds can be deceived by appearances because, in cases of fallacious appearances, they are affected *in exactly the same way* as in cases of veridical appearances. So, what is common, in the overall experiential state of a subject, in all cases of veridical appearances of something to a subject, must also be present in the overall experiential state of a subject undergoing a fallacious appearance of the same thing. What explains the necessity of (5), from the point of view of our naïve theory of mind/epistemology, is, roughly, the idea that if a mind was affected in a *different way* in a case of fallacious and veridical appearances of the same thing, it would be in a position to distinguish, on the sole basis of the appearances, between fallacious and veridical appearances of the same thing; but then 'deceptive appearances' could not exist, at least not in the full and simple sense used by our naïve theory of mind/epistemology.

as innate and modular, do not really consist in statements, but in sets of capacities and inferential mechanisms. However, our naïve theory of mind/epistemology could very well be a series of capacities and inferential mechanisms which function in such a way that it would correspond to these statements. Actually, I think that it is rather plausible, given that these statements themselves seem intuitively quite plausible.[9,10]

According to the TIH, phenomenal introspection is deeply theoretically informed by these theories. Therefore, when we introspect phenomenal states, we represent ourselves as being in states that are *experiences*, in the sense described by the theory: states in virtue of which certain states of affairs appear to us, which are gifted with special mental and epistemological properties, so that they satisfy the definition of *experiences* according to the theory — they are represented as satisfying (1–5), in virtue of the functional/inferential role of the relevant representations.[11]

5. How the TIH can solve the hardest aspect of the illusion problem

Let us suppose that the TIH is true, and that our introspective processes and representations are theoretically determined in the way I just described. Now, let us ask the question: what happens when we think intuitively — that is to say, within the framework of our naïve theory of mind/epistemology — that someone is undergoing an

---

[9] This is not to say, of course, that every philosophical theory endorses these statements. Disjunctivists, for example, notably reject (5), and (4) could probably be questioned by some contemporary theories of consciousness.

[10] It should be noted that there is a 'naïve' and intuitive concept of *appearance* that does not correspond to the concept of appearance that I just described as being the one used by our naïve theory of mind/epistemology. For example, we have a purely epistemic concept of appearance, such as the one we express linguistically when we say 'Einstein's theory of relativity appears to be true'. I think that this concept can be ultimately defined by our naïve theory of mind/epistemology in terms of phenomenal appearances and beliefs, but I suggest simply setting aside this issue, which is beyond the scope of this paper.

[11] It should be noted that, from my point of view, given that (1–5) determine the content of our concepts of phenomenal states, anyone who claims that these five points do not characterize accurately our mental functioning cannot be a full-blown phenomenal realist, and is necessarily committed (though maybe only implicitly) to a form of illusionism. Of course, given that this commitment would be merely implicit, we should certainly not expect every philosopher who denies that the functioning of our mind satisfies (1–5) to explicitly endorse illusionism.

illusion? We think that this person has a fallacious appearance of something. In most cases, this creates no problem: there can be fallacious appearances of many things, and therefore many things can be merely *illusory*, even though we don't necessarily *believe* that they are illusory. For example, let us take the example of an appearance of a red rose. We think that a red rose (or, more precisely, the presence of a red rose) can appear to a subject. Let us take all the cases in which a red rose appears veridically to a subject, and consider what is common to all these cases, regarding the way in which the subject is affected: this common element is an *experience* of a red rose. And when we think that someone is undergoing an *illusion* of a red rose, we think that this person is having an experience of a red rose (she is affected in exactly the same way as she is affected in all the cases of veridical appearances of a red rose), except that there is no red rose. So far, so good.

However, in the case of consciousness, and *only* in the case of consciousness, something troubling happens. When we try to think intuitively — using the mental resources of our naïve theory of mind/epistemology — that our current experience of a red rose, for example, is purely illusory, and that we in fact have no experience of a red rose but merely a *fallacious appearance* of an experience of a red rose, we encounter a peculiar difficulty. In order to see why, let us consider what is common, regarding the way in which subjects are affected, to all the cases of *veridical appearances* of an *experience* of a red rose. This common element would be an *experience* of an experience of a red rose, so to speak — a second-order experience. But, given the principles of our naïve theory of mind/epistemology, this state will crucially *include* the first-order experience it is an experience of[12] — it will include an experience of a red rose. Indeed, an experience of a red rose is *part* of what is common, regarding the way in which subjects are affected, to all cases of *veridical appearances of an experience of a red rose*.[13] But now what happens when

---

[12] Many philosophers argue that experiences of experiences not only *include* first-order experiences, but are actually *nothing but first-order experiences*, so that there are *no second-order experiences* distinct from first-order experiences. This is sometimes remarked by people who insist that there is no specific 'phenomenology of introspection' (Lycan, 1996; Shoemaker, 1994; Siewert, 2012). I will not say more about this kind of thesis here, as it is beyond the scope of this paper.

[13] Indeed, in all the situations in which it appears veridically to a subject that she has an experience of a red rose, (1) she will have an experience of a red rose (by definition of

we try to think intuitively that we have a fallacious appearance of an experience of a red rose? We try to think that we are in a state which is an experience of an experience of a red rose, without having an experience of a red rose. But this is impossible, as the experience of an experience of a red rose has to include an experience of a red rose.

Let's sum that up: when we try to think intuitively that we have a fallacious appearance of a given experience, we try to think that we are affected *in exactly the same way* as we are affected *in all the cases in which it veridically appears to us* that we have this given experience. But, in all these cases of veridical appearance, the way in which we are affected crucially includes the fact that we have this given experience. So, we cannot be affected in a way that is exactly similar to cases of veridical appearances of the experience *without* having this given experience — that would lead to an obvious contradiction, given the content of the representations provided by our naïve theory of mind/epistemology.

So, what the TIH predicts is that, given the nature of the naïve theory of mind/epistemology that determines our introspective representations, we are unable to think of second-order experiences that do not *include* the first-order experiences they are experiences of. Therefore, we are unable to think of a fallacious appearance of an experience. This peculiar difficulty arises in the case of consciousness, and *only* in the case of consciousness. For any other thing, we have no difficulty thinking of a *fallacious appearance* of this thing.

If the TIH is true, then, it means that, to the extent that we use an *intuitive* and *naïve* concept of illusion (fallacious appearance), an '*illusory experience*' is a contradiction. This explains the unique intuitive resistance we face when we try to accept illusionism, given that it explains why we cannot fully make sense of the idea that our experiences are purely illusory, while we don't have this problem in the case of any other thing.

Of course, in this view, illusionism is still a perfectly *coherent* hypothesis. But we need to use another concept of illusion to make it

---

the situation, which is a situation of *veridical* appearance of such an experience); (2) this experience of a red rose will be part of the way in which subjects are affected (by definition of what an experience is according to our naïve theory of mind/epistemology). So, if we consider what is common, regarding the way in which a subject is affected, to all the situations in which it appears veridically to the subject that she has an experience of a red rose, we will find that this common element necessarily includes an experience of a red rose.

completely intelligible — not the naïve concept of illusion, provided by our naïve theory of mind/epistemology, but rather a functional concept of illusion. For example, we have to use a concept of illusion that states that an illusion simply consists in, say, a state that grounds an ongoing, systematic, and cognitively impenetrable disposition to believe something false. When we use such a concept of illusion, illusionism regarding consciousness is not problematic at all, and it is quite easy to understand and to make sense of.[14]

The TIH therefore explains why denying the existence of genuine phenomenal states is 'the sort of thing that can only be done by a philosopher' (Chalmers, 1996, p. 188): we can coherently think that consciousness is an illusion only if we use a highly abstract and theoretical concept of illusion. But, as soon as we try to intuitively make sense of this idea, we are back using our naïve and innate concept of illusion, and we cannot help but find illusionism about consciousness deeply puzzling and barely intelligible. The moment we turn away from the abstract, objective, and impersonal understanding of illusionism about consciousness (in which the illusion of phenomenality is defined in purely functional terms), and try to consider what illusionism 'amounts to' or 'means' in an intuitive manner, we find this position to be unacceptable, preposterous, and hardly intelligible. However, it is crucial to note that this obviously does *not* prevent illusionism (correctly understood) from being true.

The TIH also explains why one very common objection to illusionism regarding consciousness relies on the idea that, in the case of consciousness, there is no appearance–reality gap, and that therefore there can be no fallacious appearance of consciousness. Frankish (this issue, p. 32) is right to note that, in order for such an objection to succeed, it has to presuppose a conception of *appearances* that begs the question against illusionism. However, the TIH explains why this objection

---

[14] It is also interesting to note that, according to my view, the thesis according to which *phenomenal states* do not exist, taken in itself (and if we don't add the idea that these states nevertheless *appear* to exist), is also not problematic at all — we find it very understandable and intelligible, even if it contradicts our beliefs. This is quite obviously the case, as most of us, for example, have no difficulty conceiving of a possible world in which there are no phenomenal states at all. What is problematic is the idea that phenomenal states do not exist *even if they appear to us to exist* (i.e. given our epistemic situation). This last idea, according to my view, is deeply problematic in virtue of the fact that the content of our theory of mind makes it so that we intuitively conceive of what appears to us in a way that implies that phenomenal states cannot appear fallaciously.

THE HARDEST ASPECT OF THE ILLUSION PROBLEM 137

nevertheless has some intuitive weight: it is because this question-begging conception of *appearances* happens to be our naïve, innate, and intuitive conception of appearances, provided by our naïve theory of mind/epistemology.

Overall, I think that the TIH is able to give a solution to the hardest aspect of the illusion problem. The fact that illusionism about consciousness is intuitively unacceptable and almost senseless is a by-product of the simplistic structure of our naïve theory of mind/epistemology, and of the manner in which this theory deeply determines both the content of our introspective representations of phenomenal states and the way we intuitively conceive of fallacious appearances of things. Illusionism about consciousness, in order to be acceptable, has to be understood while using a highly abstract and theoretical conception of 'illusions' and 'appearances'. As soon as we seek an *intuitive* grasp of illusionism, this position strikes us as absurd.[15]

6. Conclusion

I think that the hardest aspect of the illusion problem — i.e. the explanation of the unique intuitive resistance we experience when facing illusionism about consciousness — cannot be solved if we understand the origin of the illusion of phenomenality simply on the model of other classical, perceptual illusions. This is why current illusionist theories of consciousness fail to solve this problem. However, I think that it is possible to solve it, if we recognize the peculiarity of introspective representations of phenomenal states. The TIH is an hypothesis that attempts to do so.

To speak broadly and metaphorically, this hypothesis states the following. Our naïve theory of mind/epistemology provides a mental tool to think about appearances (and potentially deceiving

---

[15] I claim that the TIH can solve the hardest aspect of the illusion problem. One could then wonder if it is sufficient to solve the full *illusion problem* by itself, or if we have to use additional claims regarding phenomenal introspection. Can we explain why it seems to us that we are phenomenally conscious even though we are not, simply by positing that our introspective representations have the kind of content they have according to the TIH? If we say 'yes', we claim that we can explain all the peculiar features that introspection presents phenomenal states as having (features that nothing has in reality) simply by the fact that our introspective representations have the content described by the TIH. I think that it is in fact possible to give such an explanation, but it would take a long argument to make a proper case for that claim. I therefore choose to set aside this more general issue, and to focus merely on the hardest aspect of the illusion problem.

appearances). This mental tool consists in our capacity to represent *experiences*, as ways in which minds are affected. However, this mental tool happens to be quite rough and rudimentary. Because of the rough representational structure of this mental tool, a problem arises when we try to apply this mental tool to itself, so to speak — that is to say, when we try to distinguish between *experiences* and *appearances of experiences that do not include these experiences*. Such a distinction is impossible, because of the simplistic structure of this mental tool. Therefore, we end up being incapable of thinking intuitively of a *fallacious appearance of a given experience*. This explains why illusionism about consciousness can never be intuitively acceptable or even intuitively fully intelligible, while we have no difficulty making intuitive sense of illusionism regarding *any other entity*. Of course, the TIH implies that our naïve theory of mind/ epistemology is rough and simplistic, but this should not come out as a surprise. After all, all our other naïve theories are simplistic, and widely inaccurate.[16] In fact, it would be rather surprising if one of our naïve, innate theories happened to be perfectly accurate and satisfying.

The TIH may have the details of the story wrong, but I think that the main idea behind the hypothesis is correct: if we want to account for the unique strength of the illusion of phenomenality, we have to understand that introspection represents phenomenal states in a way that characterizes them as having special epistemological properties, and that in turn accounts for the difficulty we face when we try to give the 'illusion treatment' to phenomenal states — when we try to think of them as purely illusory. My guess is that a fully satisfying illusionist theory of consciousness would have to say something in the vicinity in order to solve the hardest aspect of the illusion problem.[17]

References

Chalmers, D.J. (1996) *The Conscious Mind: In Search of a Fundamental Theory*, New York: Oxford University Press.

---

[16] Our naïve physics, for example, is full or errors and approximations concerning the features of physical objects. It notably contains the idea that there are basic forces such as 'sucking' (used, for example, in a vacuum cleaner), of that the movement of bodies is caused by a kind of 'impetus', etc. (Hayes, 1978; McCloskey, 1983). The same could be said of our naïve biology.

[17] I would like to thank Samuel Webb, Fabien Mikol, Keith Frankish, David Chalmers, and two anonymous referees for their comments and their help.

Chalmers, D.J. (2006) Perception and the fall from Eden, in Szabo, T. & Hawthorne, J. (eds.) *Perceptual Experience*, pp. 49–125, Oxford: Oxford University Press.
Frances, B. (2008) Live skeptical hypotheses, in Greco, J. (ed.) *The Oxford Handbook of Skepticism*, pp. 225–244, Oxford: Oxford University Press.
Frankish, K. (this issue) Illusionism as a theory of consciousness, *Journal of Consciousness Studies*, **23** (11–12).
Graziano, M. (2013) *Consciousness and the Social Brain*, Oxford: Oxford University Press.
Hayes, P. (1978) The naive physics manifesto, in Michie, D. (ed.) *Expert Systems in the Micro-Electronic Age*, Edinburgh: Edinburgh University Press.
Humphrey, N. (2011) *Soul Dust: The Magic of Consciousness*, Princeton, NJ: Princeton University Press.
Kammerer, F. (2016) Conscious experiences as ultimate seemings: Renewing the phenomenal concept strategy, *Argumenta*, **1** (2), pp. 233–243.
Lycan, W. (1996) *Consciousness and Experience*, Cambridge, MA: MIT Press.
McCloskey, M. (1983) Naïve theories of motion, in Gentner, D. & Stevens, A. (eds.) *Mental Models*, Mahwah, NJ: Lawrence Erlbaum Associates.
Nichols, S. & Stich, S. (2003) How to read your own mind: A cognitive theory of self-consciousness, in Smith, Q. & Jokic, A. (eds.) *Consciousness: New Philosophical Perspectives*, Oxford: Oxford University Press.
Pereboom, D. (2011) *Consciousness and the Prospects of Physicalism*, Oxford: Oxford University Press.
Searle, J. (1997) *The Mystery of Consciousness*, New York: The New York Review of Books.
Shoemaker, S. (1994) The phenomenal character of experience, *Philosophy and Phenomenological Research*, **54** (2), pp. 291–314.
Siewert, C. (2012) On the phenomenology of introspection, in Smithies, D. & Stoljar, D. (ed.) *Introspection and Consciousness*, Oxford: Oxford University Press.
Strawson, G. (1994) *Mental Reality*, Cambridge, MA: MIT Press.

Pete Mandik

Meta-Illusionism and Qualia Quietism

Abstract: *Many so-called problems in contemporary philosophy of mind depend for their expression on a collection of inter-defined technical terms, a few of which are* qualia, phenomenal property, *and* what-it's-like-ness. *I express my scepticism about Keith Frankish's illusionism, the view that people are generally subject to a systematic illusion that any properties are phenomenal, and scout the relative merits of two alternatives to Frankish's illusionism. The first is phenomenal meta-illusionism, the view that illusionists such as Frankish, in holding their view, are themselves thereby under an illusion. The second is qualia quietism, the view that nothing worth saying is said by employing any of the aforementioned inter-defined technical terms.*

I am strongly sympathetic with Frankish's illusionism, and if he were right to suggest that the only real options regarding so-called phenomenal consciousness were radical realism about phenomenal properties and his illusionism, I would readily ally with the latter. However, I don't view those two as the only appealing choices, and I don't think the other views Frankish mentions prior to narrowing the field — e.g. conservative realism, eliminativism — adequately describe my own favoured approach to the topic.

Taking inspiration from Rey's (2007) meta-atheism, which, instead of the view that God does not exist, is the view that no one actually believes that God exists (despite their claims to the contrary), I am tempted to label my reaction to Frankish's illusionism, *meta-illusionism*. The gist of meta-illusionism is that it rejects phenomenal realism while also insisting that no one is actually under the illusion

Correspondence:
Pete Mandik, William Paterson University, Wayne, NJ, USA.
Email: MandikP@wpunj.edu

that there are so-called phenomenal properties. One line of thought that leads me to meta-illusionism is that there's no content to the claim that any properties seem 'phenomenal' to anybody. And here my worries are almost entirely about the word 'phenomenal' and its ilk. Another line of thought that leads me to meta-illusionism hinges on the word 'illusion' and the worry that, while there may be some people to whom it seems that there are phenomenal properties, this appearance, which I do not grant as reflecting reality, is insufficiently widespread to be worth considering an *illusion* (as opposed to, say, a weird belief that certain philosophically educated individuals claim to hold). Given my worries about what counts as an illusion properly so-called, I am a little reluctant to embrace the label 'meta-illusionism' since I don't want to say that the illusionists are deceived in a sufficiently systematic way to attribute to them an illusion. If I needed a better label for my view, I might go with 'qualia quietism', and I'll close the current article with remarks about what that amounts to.

1. On 'phenomenal'

One thing Frankish and I have in common is that neither of us wants to assert that there are any properties instantiated that are referred to or picked out by the phrase 'phenomenal properties'. One place where Frankish and I part ways is over whether that phrase is sufficiently meaningful for there to be a worthwhile research programme investigating how it comes to seem to people that their experiences instantiate any such properties. Like Frankish, I'm happy with terms like 'experience', 'consciousness', and 'conscious experience' and join Frankish in using what he calls 'weak' and functional construals of such terms. But, unlike Frankish, I see no use at all, not even an illusionist one, for the term 'phenomenal' and its ilk.

The term 'phenomenal', as used in contemporary philosophy of mind, is a technical term. I am aware of no non-technical English word or phrase that is accepted as its direct analogue. Unlike technical terms in maths and physics, which are introduced with explicit definitions, 'phenomenal' has no such definition. What we find instead of an explicit definition are other technical terms treated as interchangeable synonyms. Frankish follows common practice in philosophy of mind when he treats 'phenomenal' as interchangeable with, for instance, 'qualitative' or, in scare-quotes, '"feely"'.

As used in the relevant philosophy-of-mind contexts, 'qualitative' doesn't mean simply 'relating to qualities', since properties and qualities are one and the same, and the technical term is supposed to pick out some special kind of property, a property distinctive of conscious experiences. The technical term 'qualitative' also does not seem to mean 'characterizing in ways other than quantity', which, if it did, would at least have the virtue of relating it to a non-technical use, but would lose its claim to pick out something specific to conscious experience.

Trouble arises for "feely", which, as the scare quotes seem to warn, is not to be equated with non-technical uses of 'feel' and its cognates, which pertain to feeling temperatures and textures, but decidedly not to, for example, what differentiates seeing blue from seeing green. These three terms — 'phenomenal', 'qualitative', and '"feely"' — form a tight circle conveying little to no information to the meta-illusionist demanding to know what it is that the illusionist thinks people are under the illusion of.

One phrase that might seem to break us out of the circle of technical terms is the phrase 'something it's like', for there are non-technical uses of 'what it's like' (Farrell, 2016), and phenomenal properties are supposed to be those properties in virtue of which there is something it's like to have experiences. However, to my knowledge, the syntactic transformation from 'what it's like' to 'there is something it's like' occurs only in technical philosophy-of-mind contexts. This makes me doubt that non-technical uses of 'what it's like', which sometimes (but not always) are employed to pick out mental states, are employed to pick out a peculiar kind of property of mental states. When, for example, pop stars sing about knowing what it's like to fall in love, they give little evidence of attributing so-called 'phenomenal' properties, as opposed to whatever other properties a meta-illusionist can readily grant are seemingly instantiated by love states. The hyphenated 'what-it's-like' in Frankish's '"what-it's-like" properties' (Frankish, this issue, p. 15) is yet another technical term shedding no light on the term 'phenomenal'.

We have then, in place of an explicit definition of 'phenomenal properties', a circular chain of interchangeable technical terms — a chain with very few links, and little to relate those links to non-technical terminology. The circle, then, is vicious. I'm sceptical that any properties seem 'phenomenal' to anyone because this vicious circle gives me very little idea what seeming 'phenomenal' would be.

One way out of the vicious circle that seems unavailable to Frankish is a kind of deferential ostension. Radical realists can, without pangs of conscience, attempt to convey what they're talking about by inviting an act of inner ostension: they invite us to look inward and appreciate that they are talking about properties like *that*. Since they think there are such properties, they aren't failing by their own lights to specify what they are talking about. They may nonetheless be failing, and their 'that' refers to nothing, but they would not be failing *by their own lights*. Illusionists, in contrast, are in a worse position, since by their own lights there are no such instantiated properties to inwardly point at. Since they hold instantiated phenomenal properties to be merely notional, they may attempt to gesture toward what they would be if there were any by some indirect route, perhaps via the description, 'whatever it is that radical realists are talking about'. However, the very real possibility arises that the radical realists *aren't* all talking about the same thing, given the lack of any explicit definition of 'phenomenal property' that they agree on. Echoing Wittgenstein (1953), each radical realist labels the private contents of their beetle-box 'a beetle', but, for all anyone knows, each box may contain something different from the others, or even nothing at all.

The problem I pose for Frankish is a problem of content, but it is not the problem of content that he himself addresses in Section 3.4 of the target article. There he considers the problem of what would fix the content of the introspective representations of phenomenal properties given that there aren't any phenomenal properties instantiated. He appreciates that the illusionist will not be able to employ any psycho-semantics that relies on positing relations, causal or other, between representations and actually instantiated properties accurately represented. The challenge I pose for Frankish is not the challenge of giving an account of *how* such representations get their content. The challenge I pose is instead to articulate *what* the content is. The contrast between the two challenges might be illustrated via the following analogy: the theory-of-content challenge is to explain *how* a representation of cows comes to be a representation of cows and not of horses and not of the disjunction cow-or-horse, and so on. My challenge — the articulation challenge — is to say *what* some target representation represents, to say that it represents cows, or instead represents horses, or some disjunction, and so on. I want to ask of the representations that Frankish alleges to be illusory: *what* are they representations of? The technical-term circle — phenomenal property, qualitative property, 'feely' property, 'what-it's-like' property, quale — does nothing to answer the question. Maybe a radical realist may feel confident that the terms in the circle are informative, picking out, as they allege, some inner ostended property — but how can one sympathetic to either illusionism or meta-illusionism gain any satisfaction here?

2. Regarding illusions

So far, I have been focusing on a circle of technical synonyms for the technical term 'phenomenal property'. To be fair, Frankish does give other characterizations of phenomenal properties, characterizations that might break us out of the circle described above. However, these other characterizations introduce other problems for Frankish's illusionism.

Frankish sometimes characterizes phenomenal properties as anomalous (p. 13) and other times as magical (p. 28). There are other characterizations that come up as well in Frankish's article (e.g. 'simple', 'intrinsic'), but I will focus on just on 'magical' and 'anomalous' for simplicity's sake. What these characterizations (including characterizations besides 'magical' and 'anomalous') have in common is that they (1) give some hope of, at least partially, breaking out of the circle of technical terms I complain about in the previous section, (2) they characterize properties in such a way that motivates illusionism along the lines spelled out in Section 2.3 of the target article, but (3) as I'll argue below, problems arise for the claim that the false appearances in question are *illusions* properly so-called.

Perhaps Frankish is not employing technical senses of either 'anomalous' or 'magical', so there's hope here of breaking out of the vicious circle of technical synonyms, and to give an informative, though perhaps partial, answer to the question of what so-called phenomenal representations are representations of. On the partial account now being scouted, they would be, if actually instantiated, properties of experiences that are anomalous or magical. The worry I want to raise now is whether it seems to people in a sufficiently systematic way that their experiences have anomalous or magical properties. If it only seems this way to a few people, people versed in certain moves internal to the philosophy of mind, then this erroneous appearance seems ill-described as an *illusion*.

Illusions properly so-called occur in systematic ways for large numbers of the population. For example, it is very easy to find people prone to the visual illusion of Müller-Lyer. For many people, certain line pairs seem of unequal length when in reality they are equal. Switching examples: people exposed to the Monty Hall problem give the wrong answer with such high frequency that it's worth, perhaps, calling it a cognitive illusion.[1]

To be clear, I am not claiming that in order for something to be an illusion properly so-called, a large number of people have to actually have undergone the illusion. Suppose very few people have seen the Müller-Lyer figures or have been exposed to the Monty Hall problem. Nonetheless, in these cases, there is an easily conveyed stimulus or scenario that, when presented to people, readily elicits a false appearance in many of those so-presented. There's something systematic to these false appearances that makes them illusions as opposed to mere false appearances. Examples of false appearances that lack this requisite systematicity include many examples of run-of-the-mill false belief. Last Wednesday, George believed his keys were in his briefcase when in reality he had left them on his desk. In some sense of appear, it appeared to George that his keys were in his briefcase. But his false representation of the keys' whereabouts is not part of a larger pattern of eliciting conditions and responses that make it rise to the level of illusion. The example of George's false belief is one that involves only a single person. However, being held by many people is not a sufficient condition for a false representation to be an illusion. Many people have held, erroneously, that 1 is a prime number and that Christopher Columbus was the first European to arrive in the Americas. But these false representations aren't illusions either. There isn't, in these cases, an easily conveyed scenario or stimulus that reliably elicits a false appearance.

Let us temporarily leave aside the topic of whether conscious experiences seem to have anomalous or magical properties, and ask about apparent anomalousness or apparent magicalness more generally. Perhaps if we have a better handle on what such appearances are generally, we may find ourselves in a better position to

---

[1] The Monty Hall problem involves a game show scenario in which there are three closed doors, behind two of which are goats, and one of which is a car. The contestant makes an initial selection of a door, but before that door is opened, one of the other doors is opened by the game show host to reveal a goat. The contestant is now offered a choice: to stick with their initial section, or to switch to the other unopened door. The central question of the Monty Hall problem is whether there's any advantage (assuming one prefers cars to goats!) to switching. The common, and wrong, answer is that there is no advantage, on the erroneous grounds that there's a 1-in-2 probability of the car being behind the initially selected door. The correct response, unintuitive to many people, is that accepting the offer to switch doors raises one's chances of getting the car from 1-in-3 to 2-in-3.

assess the claim that conscious experiences systematically appear to have anomalous or magical properties.

Consider some impressive display performed by a professional magician. Suppose a ball is presented on a table's surface, and then seemingly covered with an opaque cup. The cup is then lifted and the ball is no longer where it was last seen, and the cup appears empty. Audiences delight in such displays and it seems natural to describe them as *apparently* anomalous or magical, and such appearances arise even for audience members who don't believe that anything *actually* anomalous or magical occurred. Despite the belief that nothing out of the ordinary occurred, there is nonetheless a reliably generable *appearance* of something out of the ordinary: a ball visible in one location and then no longer being visible in that location without being seen to have been removed, and also without any visibly obvious means by which it may have been moved while unseen. The reliability with which magicians can elicit such responses from audience members, across wide varieties of audience member, makes it worth calling such inaccurate appearances illusions.

Let us return now to the way our own conscious experiences seem to us. That a wide variety of people have conscious experiences (again, in the 'weak' and functional sense of conscious experience that the illusionist will grant) is not something I'm calling into question. Nor do I intend to raise any doubts about whether those conscious experiences appear to people in various ways. For example, when someone is in a position to answer the question of whether they saw, or instead felt, that a piece of metal is hot, I don't mind chalking this up to the way their experience of the metal seemed to them. It seemed to them, for example, that they had a tactile experience as opposed to a visual experience of the metal's hotness. But what I have doubts about is whether people unfamiliar with contemporary philosophy of mind introspect their conscious experiences and find them to seemingly have properties that are anomalous or magical. Suppose one is seeing some expanse of green and also seems to oneself to be seeing an expanse of green. What anomalous or magical properties are apparently revealed to introspection? In the senses of 'anomalous' and 'magical' that we would cash out in terms of *being out of the ordinary* or *being contrary to the laws of nature*, speaking for myself, my introspected experiences never appear that way. For example, when it seems to me that I'm having a visual experience of green, it doesn't also seem to me that my experience has properties that violate the laws of nature.

Perhaps the apparent anomalousness or magicalness isn't supposed by the illusionists to arise in ordinary episodes of introspection, but instead on occasions of reflection when one contemplates the sorts of topics connected to philosophical discussions of alleged explanatory gaps between conscious experiences and neuro-functional states. Perhaps the story is supposed to go like this: one attends to, say, a visual experience of a vivid electric blue, while also contemplating the brain processes that the experience is allegedly identical to or explainable in terms of, and it's *that* — the alleged fact of the one arising out of, being identical to, or explainable in terms of the other — that seems magical or anomalous. But what's the sense of 'magical' or 'anomalous' at play here? It's unlikely to be some construal in terms of *being out of the ordinary* or *contrary to the laws of nature*. What we have here is the ordinary and lawful way in which visual experiences of vivid blue arise: they arise, ordinarily, in connection with the activities of human brains and probably also the activities in many of the brains of non-human animals. Further, and more to the point, it is not part of our introspective contents that such experiences don't arise in this manner. Our experiences do not generally present themselves to introspection (as opposed to comparatively more complex episodes involving theoretically sophisticated reflection) as the sorts of things that do not or cannot arise in virtue of neural processes.

Of course, there are many in the philosophy of mind who have wondered *how* and *why* it is that experience so arises, and some philosophers have also been tempted towards the conclusion that no satisfying explanations of the how and the why can be given. But appreciating these philosophical questions and positions, and feeling any pull one way or another regarding them, seems to require far more philosophical sophistication than sits well with any apparent anomalousness or magicalness being illusions properly so-called.

Even if the problems I tried to pose in the present section turn out to not be genuine problems for Frankish's illusionism, the problems raised in the previous section may still loom large. This is because the characterizations of phenomenal properties as anomalous or magical seem partial at best. It is unlikely that anyone who thinks that there is content to the phrase 'phenomenal properties of experience' will accept that a property of experience is phenomenal if and only if it is anomalous or magical. (I take similar remarks to apply to other characterizations of the phenomenal that Frankish employs, such as 'intrinsic'.)

3. Qualia quietism

I have expressed doubts that anyone undergoes an illusion properly so-called that experiences instantiate phenomenal properties. Some of my doubts hinge on scepticism about any representations having 'phenomenal' content. Other of my doubts hinge on reservations about what properly counts as an illusion. Given the latter sort of doubt, I'm reluctant to attribute to Frankish a meta-illusion — the illusion that anyone is undergoing the illusion of phenomenal properties. I'm likewise reluctant to embrace the label of *meta-illusionism*. More appealing to me is to recommend *qualia quietism*, which we may characterize as the view that the terms 'qualia', 'phenomenal properties', etc. lack sufficient content for anything informative to be said in either affirming or denying their existence. Affirming the existence of what? Denying the existence of what? Maintaining as illusory a representation of what? No comment. No comment. No comment.

This is not to assert that the field of consciousness studies is pure folly. As Frankish stresses, there are plenty of clear uses of 'conscious', 'experience', and 'conscious experience', uses that can be explicated in 'weak' and functional ways. And there's plenty of work already done and still being done on consciousness so-construed. But I see no future, illusionist or realist, for the term 'qualia' and its ilk.[2]

References

Farrell, J. (2016) 'What it is like' talk is not technical talk, *Journal of Consciousness Studies*, **23** (9–10), pp. 50–65.
Frankish, K. (this issue) Illusionism as a theory of consciousness, *Journal of Consciousness Studies*, **23** (11–12).
Rey, G. (2007) Meta-atheism: Religious avowal as self-deception, in Antony, L. (ed.) *Philosophers Without Gods*, Oxford: Oxford University Press.
Wittgenstein, L. (1953) *Philosophical Investigations*, Oxford: Blackwell.

---

[2] I am very grateful to Keith Frankish for the opportunity to prepare this article. I am grateful for conversations on these topics with him as well as with Richard Brown, Jonathan Farrell, David Pereplyotchik, David Rosenthal, Eric Steinhart, and Josh Weisberg.

Nicole L. Marinsek and Michael S. Gazzaniga

A Split-Brain Perspective on Illusionism

Abstract: The split-brain literature offers a unique perspective on theories of consciousness. Since both the left and right hemispheres of split-brain patients remain conscious following split-brain surgery, any theory that attempts to explain consciousness in neurotypical individuals must also be able to explain the dual consciousness of split-brain patients. This commentary examines illusionism — the theory that phenomenal properties are illusory — through the lens of the split-brain literature. Based on evidence that both hemispheres of split-brain patients are capable of introspection and both hemispheres can experience and maintain illusions, it is theoretically possible that phenomenal properties are illusions created by distorted introspection, in accordance with illusionism. However, in order to appropriately evaluate whether illusionism is a valid explanation of consciousness in split-brain patients, it is imperative that neural mechanisms are proposed that explain how introspection gives rise to illusory phenomenal properties.

Any theory that attempts to relate conscious experiences to the brain must at some point account for the unique consciousness of a split-brain patient. Split-brain patients, who have had their corpora callosa severed as a treatment-of-last-resort for severe epilepsy, are strangely normal following surgery. They maintain meaningful conversation, move about in a coordinated fashion, exhibit appropriate desires and

Correspondence:
Nicole L. Marinsek and Michael S. Gazzaniga, University of California, Santa Barbara, CA, USA.   *Email: marinsek@dyns.ucsb.edu*

emotions, maintain social relationships, and many even hold a job. Zaidel (1994) writes, 'Their walk is coordinated, their stride is purposeful, they perform old unilateral and bimanual skills, converse fluently and to the point, remember long-term events occurring before surgery, are friendly, kind, generous, and thoughtful to the people they know, have a sense of humor, and so on down a whole gamut of what it takes to be human' (pp. 9–10).

Disconnecting the two hemispheres does not noticeably impair consciousness. Instead, it appears to *split* consciousness: there is ample evidence suggesting that the two hemispheres possess independent streams of consciousness following split-brain surgery (see Marinsek, Gazzaniga and Miller, 2016). Therefore, any theory that proposes mechanisms for consciousness in neurotypical individuals must also be able to explain the preserved consciousness in each hemisphere of a split-brain patient.

In this commentary, we will examine whether the theory of illusionism is compatible with the split-brain literature. Illusionism posits that the phenomenal properties of consciousness are an illusion produced by the limitations of introspection:

> Illusionists deny that experiences have phenomenal properties and focus on explaining why they seem to have them. They typically allow that we are introspectively aware of our sensory states but argue that this awareness is partial and distorted, leading us to misrepresent the states as having phenomenal properties. (Frankish, this issue, p. 14)

In order to determine whether illusionism can account for consciousness in split-brain patients, we will address the following three questions:

1. Does each hemisphere of a split-brain patient have phenomenal experiences?
2. Is each hemisphere of a split-brain patient capable of introspection?
3. Is there any evidence that each hemisphere of split-brain patient can create and maintain an illusion?

As we attempt to answer these questions, we may ascribe thoughts, intentions, and behaviours to the left and right hemispheres of split-brain patients. We do so out of convenience and not to anthropomorphize the hemispheres. When we refer to the disconnected left hemisphere, please note that it is more accurate to refer to the entire split-brain patient, minus the disconnected right hemisphere, and vice versa for the right hemisphere.

1. Do split-brain patients have phenomenal experiences?

In order to determine if illusionism holds for split-brain patients, we must first verify that split-brain patients experience the subjective feelings associated with phenomenal consciousness. Our aim is to determine whether split-brain patients experience the phenomenal properties that neurotypical individuals experience, regardless of whether those properties are real or illusory. Therefore, when we refer to 'phenomenal consciousness', we are referring to conscious experiences that either have phenomenal properties or simply appear to have them. According to Block (2005), phenomenal consciousness refers to the *content* of an experience: 'phenomenally conscious content is what differs between experiences as of red and green' (p. 46). Put a different way, Block also states, 'Phenomenal consciousness is experience; the phenomenally conscious aspect of a state is what it is like to be in that state' (1995, p. 227). Phenomenal consciousness differs from access consciousness, which refers to the accessibility of information for processing by cognitive modules or networks (Block, 2005). Access consciousness does not refer to the content of an experience, but rather its availability for use by different cognitive modules, such as those that support memory, decision making, reasoning, action, and so on.

There is little doubt that the left hemisphere is phenomenally conscious. Split-brain patients report that they feel no different following their surgeries. Since the neural machinery supporting language is lateralized to the left hemisphere, we can attribute patients' self-reports to the left hemisphere. Based on the verbal reports of the left hemisphere, there is no indication that the subjective consciousness of the left hemisphere is any different from the subjective consciousness of neurotypical individuals, and we can presume that the left hemisphere possesses both access consciousness and phenomenal consciousness.

Characterizing the conscious experience of the disconnected (and mute) right hemisphere is more challenging. Because the right hemisphere has an impoverished language system, we cannot rely on verbal introspective reports to determine that it is conscious. Instead, we must infer the conscious status of the right hemisphere based on its capabilities and behaviours in controlled experiments.

When we examine the behaviours and abilities of the disconnected right hemisphere, it is clear that it has access consciousness. Block states that access conscious content is comprised of 'information about which is made available to the brain's "consumer" systems: systems of memory, perceptual categorization, reasoning, planning, evaluation of alternatives, decision-making, voluntary direction of attention, and more generally, rational control of action' (2005, p. 47). In controlled experiments, the right hemisphere is able to independently remember information (Phelps and Gazzaniga, 1992; Metcalfe, Funnell and Gazzaniga, 1995), categorize objects (Metcalfe, Funnell and Gazzaniga, 1995), make inferences about perceptual causality (Roser *et al.*, 2005), make predictions (Wolford, Miller and Gazzaniga, 2000), control attention voluntarily (Holtzman *et al.*, 1981), and initiate purposeful movement (reviewed in Gazzaniga, 2000). The fact that the right hemisphere's 'consumer' systems are functionally intact indicates that relevant information is accessible within the right hemisphere, and suggests that the disconnected right hemisphere meets the criteria for possessing access consciousness.

It is less certain whether the right hemisphere is *phenomenally* conscious, in the sense that it experiences phenomenal feelings. No split-brain studies to our knowledge have set out to determine whether or not the right hemisphere is phenomenally conscious, and it is unclear whether it is even possible to do so without relying on verbal or written self-report. We do know that the disconnected right hemisphere can distinguish between contents that have different phenomenal properties (to neurotypical individuals): it can distinguish between different colours (Wolford, Miller and Gazzaniga, 2000), different sounds (Musiek, Pinheiro and Wilson, 1980), and different touch sensations (Zaidel, 1998), for example. Although we cannot know for certain whether these different sensations are associated with different phenomenal feels, we also cannot know for certain that they are not. Block makes the argument that introspective self-reports are not necessary to verify phenomenal consciousness: 'you don't need reports about the subject's experiences to get good evidence about what the subject is experiencing: indications of what the subject takes to be in front of him will do just fine' (2005, p. 51). Based on the evidence that the right hemisphere has sensory experiences and meets the criteria for access consciousness, we can presume that it also has phenomenal consciousness even though it cannot describe the contents or subjective feelings of its experiences.

2. Are split-brain patients capable of introspection?

The second question we need to address is whether or not the disconnected hemispheres of split-brain patients are capable of introspection. According to illusionism, phenomenal feelings arise because introspection misrepresents sensory states. Frankish argues that — just like how our visual systems create the illusion that objects are coloured — our introspective systems may create the illusion that sensations have phenomenal feels:

> Sensory states have complex chemical and biological properties, representational content, and cognitive, motivational, and emotional effects. We can introspectively recognize these states when they occur in us, but introspection doesn't represent all their detail. Rather, it bundles it all together, representing it as a simple, intrinsic phenomenal feel. Applying the magic metaphor, we might say that introspection sees the complex sleight-of-hand performed by our sensory systems as a simple magical *effect*. (Frankish, this issue, p. 18)

In order for illusionism to explain the phenomenal consciousness of each hemisphere in a split-brain patient, there must be some evidence that the hemispheres are independently capable of introspection. The split-brain literature provides several examples of such evidence.

In 1977, LeDoux, Wilson and Gazzaniga conducted a series of experiments that assessed the introspective abilities of the left and right hemisphere of a split-brain patient (patient P.S.). In the first study, a word was presented to the patient's left or right hemisphere and the patient judged how good or bad the word was by pointing to a 7-point Likert scale, where 1 represented good and 7 represented bad. The left and right hemisphere reported similar ratings for a few of the words; for example, both the left and right hemispheres rated *car* and *money* as good (1) and *vomit* as somewhat bad (5). However, most of the hemispheres' ratings differed substantially. Six out of the twelve words were associated with at least a 4-point rating gap, with the right hemisphere almost always giving a more negative rating. For example, the left hemisphere rated the words *nice*, *mother*, *sex*, and *Paul* (the patient's own name) as good (1), but the right hemisphere rated these words as bad (6–7).

In a second study from the same series, LeDoux, Wilson and Gazzaniga (1977) presented a word to the left or right hemisphere and asked the patient to indicate how much he liked the word by pointing to one of five options, ranging from 'like very much' to 'dislike very much'. This time, the ratings of the left and right hemispheres were quite consistent: the left and right hemispheres gave 12 of the 16 words the same rating, and the rating gaps of words with different ratings were generally much smaller than in experiment 1. Not only were the ratings consistent, they were also quite reasonable and did not appear to be random or meaningless. The left and right hemisphere reported that they 'very much liked' the words *home*, *church*, *mom*, *dad*, *Paul*, and *Fonz* (the TV character) and 'liked' the words *sex*, *school*, *police,* and *Liz* (the patient's girlfriend). The only word that received consistently poor ratings was *Nixon*, which the right hemisphere reported it 'disliked' and the left hemisphere reported it 'disliked very much'.

In the third experiment, the researchers presented a question to the right hemisphere and asked the split-brain patient to spell out his answer using Scrabble tiles. When asked 'Who are you?' the patient correctly spelled out PAUL. When asked who his favourite girl was, the patient spelled out LIZ, his girlfriend. When asked what his mood was, the right hemisphere spelled GOOD and then SILLY when asked again later. Finally, when asked what job he wanted, the patient spelled AUTOMOBILE RACE, even though the patient routinely said (via the left hemisphere) that he wanted to be a draftsman.

In another experiment, Sperry, Zaidel and Zaidel (1979) showed pictures to the left or right hemisphere of a different split-brain patient (L.B.) and asked him to give the picture a thumbs-up or a thumbs-down based on how he felt about it. The ratings of the right hemisphere were identical to the left hemisphere's verbal reports and were again quite reasonable: 'LB had responded with "thumbs-down" evaluations for Castro, Hitler, overweight women in swim suits, and a war scene. Intermixed with these and other responses, "thumbs-up" signals were obtained for Churchill, Johnny Carson, pretty girls, scenes from ballet and modern dance and a horizontal neutral thumb signal for Nixon' (*ibid.*, p. 163).

Taken together, these experiments demonstrate two important points. First, the left and right hemispheres are both capable of accessing their mood, desires, feelings, and opinions about things, people, and themselves. It is important to note that it is possible that the right hemisphere's responses reflect conditioned associations rather than purposeful introspection. That is, it is possible, for example, that the right hemisphere gave Hitler a thumbs-down because it has many low-level, negative associations with Hitler and not because it introspectively accessed its feelings when Hitler's picture was presented. However, the right hemisphere's ability to report its opinions and desires by arranging Scrabble tiles, which have an infinite number of possible arrangements, suggests that the reports of the right hemisphere extend beyond simple conditioned associations and instead reflect true introspection. Second, these experiments demonstrate that the opinions of the left and right hemisphere are independent — sometimes they agree and sometimes they do not. This provides further evidence that the conscious experiences of the left and right hemispheres are distinct and independent, and illusionism (or any other theory of consciousness) must be able to account for both.

3. Do split-brain patients experience illusions?

Frankish makes the argument that phenomenal feelings are a special type of illusion, similar to the illusion of colour or the illusion of continuous motion in film or cartoons. What separates phenomenal feelings from other types of illusions is that they are inherently subjective and can only be observed from one vantage point (that is, via introspection). We cannot directly determine whether each hemisphere is capable of creating the illusion of phenomenal feelings, but we can explore whether the hemispheres are capable of creating and maintaining other sorts of illusions. If we find evidence that both hemispheres of a split-brain patient experience a variety of illusions, it may be more likely that the hemispheres are capable of creating the illusion of phenomenal feelings in accordance with illusionism.

Evidence suggests that both the left and right hemispheres of split-brain patients experience perceptual illusions. For example, both hemispheres have been shown to perceive motion when there is none (Corballis *et al.*, 2004) and perceive contours where there are none (Corballis *et al.*, 1999), and the right hemisphere has been shown to judge that the trajectories of two colliding shapes are causal (Roser *et al.*, 2005).

There is also evidence that the left hemisphere experiences an illusion of control. The illusion of control refers to instances when the left hemisphere mistakenly claims ownership of an action that was actually initiated and carried out by the right hemisphere. Marinsek, Gazzaniga and Miller (2016) describe a case where the command to stand up was presented to the right hemisphere of a split-brain patient. After the patient stood, the experimenters asked him why he stood up. Even though the left hemisphere was not given the command to stand, the patient (speaking with his left hemisphere) explained that he stood up because he was thirsty and wanted to get a drink. The left hemisphere not only claimed ownership of the act, but it also ascribed an intention to it. By doing so, the left hemisphere maintained the illusion of control over the body. In a similar example, the word 'smile' was presented to the right hemisphere and the word 'face' was presented to the left hemisphere and the patient was asked to draw what he saw. The patient drew a smiley face and the experimenter asked him why he did so. Again, the patient's left hemisphere offered an explanation that maintained the illusion of control, saying 'What do you want, a sad face? Who wants a sad face around?' (Gazzaniga, 2013, p. 14).

Closely related to the illusion of control is the illusion of unity, and it is perhaps the most striking feature of split-brain patients. As we have said, split-brain patients do not *feel* disconnected or disunified following their surgeries. The patients feel unified even though they have split brains and, by all indications, split minds (see Marinsek, Gazzaniga and Miller, 2016). The continuation of the patients' subjective feelings of unity is evident in the responses of some split-brain patients during testing. For example, Sperry, Zaidel and Zaidel (1979) asked patient L.B.'s right hemisphere to give a thumbs-up or a thumbs-down to various pictures, and at some point during the testing the right hemisphere gave a thumbs-down to three pictures in a row. When the experimenter questioned the third thumbs-down, the patient's left hemisphere remarked, 'Guess I'm antisocial' (*ibid.*, p. 160). Not only did the left hemisphere assume responsibility for the right hemisphere's negative responding, but it implied that the right hemisphere's behaviour was reflective of its own self-identity, and not that of some other entity. That is, the left hemisphere didn't say, 'my right brain is acting antisocial' or 'I wasn't the one who gave the thumbs-down', as might be expected if there was no illusion of unity.

In some cases, the left and right hemispheres have conflicting intentions and the left and right hands attempt to carry out different actions. Interestingly, when conflict arises, the illusion of control breaks down but the illusion of unity still holds. For example, one of the first split-brain patients reported: 'The muscles of my left side do not coordinate very well with those of the rest of my body. For instance, I find myself trying to open a door with the right hand and at the same time trying to push it shut with the left; putting my dress on with the right and pulling it off with the left' (Van Wagenen and Herren, 1940, p. 756). In this example, the patient (speaking with her left hemisphere) suggested that she lacks full control of her left hand (which is largely controlled by the right hemisphere), but she attributed her lack of coordination to the muscles in her left side and not to some other mind or intentional agent. The split-brain literature contains many more instances where patients indicate that they are not in control of their left hands, but the patients never give any indication that they possess two duelling minds or that their conscious experience is fragmented or split in any way (Zaidel, 1994).

The illusions of control and unity may be maintained by the left hemisphere interpreter, a cognitive module rooted in the left hemisphere that creates causal explanations (Gazzaniga, 1989). The effects of the interpreter are apparent when the left hemisphere is asked to explain the right hemisphere's behaviour. Even though the left hemisphere does not have access to the information presented to the right hemisphere, or the thoughts, intentions, and desires of the right hemisphere, the interpreter will offer an explanation for the right hemisphere's behaviour. The explanations of the interpreter are often plausible — for example, the patient's explanation that he stood up to get a drink sounds completely rational if you did not know that the right hemisphere was given the command to stand. More importantly, the interpreter's rationalizations help maintain the illusions of unity and control. As Gazzaniga writes:

> The interpreter is driven to generate explanations and hypotheses regardless of circumstances. The left hemisphere of split-brain patients does not hesitate to offer explanations for behaviours, which are generated by the right hemisphere. In neurologically intact individuals, the interpreter does not hesitate to generate spurious explanations for sympathetic arousal. In these ways, the left hemisphere interpreter may generate a feeling in all of us that we are integrated and unified. (2000, p. 1319)

The effects of the interpreter are more visible in split-brain patients where we know there is a lack of control and unity, but it likely contributes to the subjective feelings of control and unity in neurotypical individuals as well.

4. Conclusion

One major limitation of the theory of illusionism is that it does not offer any mechanisms for how the illusion of phenomenal feelings works. As anyone who has seen a magic trick knows, it's quite easy to say that the trick is an illusion and not the result of magical forces. It is much, much harder to explain how the illusion was created. Illusionism can be a useful theory if mechanisms are put forth that explain how the brain creates an illusion of phenomenal feelings. When mechanisms are proposed, the split-brain literature will be an important testing ground for determining whether the candidate mechanisms can account for the split consciousness of split-brain patients.

One thing the split-brain literature tells us is that phenomenal consciousness may not be the product of one grand illusion. Instead, phenomenal consciousness may be the result of multiple 'modular illusions'. That is, different phenomenal feelings may arise from the limitations or distortions of different cognitive modules or networks. This idea is echoed by Block (2005) who suggests that phenomenal experiences are produced by local neural processing, such that recurrent neural activity in area MT/V5 creates the phenomenal feeling of motion and recurrent neural activity in the fusiform face area creates the experience of seeing a face. The left and right hemispheres of a split-brain patient may have different phenomenally conscious experiences because they house different specialized neural networks. It may be possible that the right hemisphere has reduced phenomenal feelings for verbal representations and the left hemisphere has reduced phenomenal feelings for visual or spatial representations. If this is the case, phenomenal consciousness can be fragmented. Illusionism therefore may not have to account for one grand illusion, but for many 'modular illusions' that each have their own neural mechanisms.

References

Block, N. (1995) On a confusion about a function of consciousness, *Brain and Behavioral Sciences*, **18** (2), pp. 227–247.

Block, N. (2005) Two neural correlates of consciousness, *Trends in Cognitive Sciences*, **9** (2), pp. 46–52.

Corballis, P.M., Fendrich, R., Shapley, R.M. & Gazzaniga, M.S. (1999) Illusory contour perception and amodal boundary completion: Evidence of a dissociation following callosotomy, *Journal of Cognitive Neuroscience*, **11** (4), pp. 459–466.

Corballis, M.C., Barnett, K.J., Fabri, M., Paggi, A. & Corballis, P.M. (2004) Hemispheric integration and differences in perception of a line-motion illusion in the divided brain, *Neuropsychologia*, **42** (13), pp. 1852–1857.

Frankish, K. (this issue) Illusionism as a theory of consciousness, *Journal of Consciousness Studies*, **23** (11–12).

Gazzaniga, M.S. (1989) Organization of the human brain, *Science*, **245**, pp. 947–952.

Gazzaniga, M.S. (2000) Cerebral specialization and interhemispheric communication: Does the corpus callosum enable the human condition?, *Brain*, **123**, pp. 1293–1326.

Gazzaniga, M.S. (2013) Shifting gears: Seeking new approaches for mind/brain mechanisms, *Annual Review of Psychology*, **64**, pp. 1–20.
Holtzman, J.D., Sidtis, J.J., Volpe, B.T., Wilson, D.H. & Gazzaniga, M.S. (1981) Dissociation of spatial information for stimulus localization and the control of attention, *Brain*, **104**, pp. 861–872.
LeDoux, J.E., Wilson, D.H. & Gazzaniga, M.S. (1977) A divided mind: Observations on the conscious properties of the separated hemispheres, *Annals of Neurology*, **2** (5), pp. 417–421.
Marinsek, N., Gazzaniga, M.S. & Miller, M.B. (2016) Split-brain, split-mind, in Laureys, S., Tononi, G. & Gosseries, O. (eds.) *The Neurology of Consciousness*, vol. 2, pp. 263–271, Amsterdam: Elsevier Ltd.
Metcalfe, J., Funnell, M. & Gazzaniga, M.S. (1995) Right-hemisphere memory superiority: Studies of a split-brain patient, *Psychological Science*, **6** (3), pp. 157–164.
Musiek, F.E., Pinheiro, M.L. & Wilson, D.H. (1980) Auditory pattern perception in 'split brain' patients, *Archives of Otolaryngology*, **106** (10), pp. 610–612.
Phelps, E.A. & Gazzaniga, M.S. (1992) Hemispheric differences in mnemonic processing: The effects of left hemipshere interpretation, *Neuropsychologia*, **30** (3), pp. 293–297.
Roser, M.E., Fugelsang, J., Dunbar, K.N., Corballis, P.M. & Gazzaniga, M.S. (2005) Dissociating processes supporting causal perception and causal inference in the brain, *Neuropsychology*, **19** (5), pp. 591–602.
Sperry, R.W., Zaidel, E. & Zaidel, D. (1979) Self recognition and social awareness in the deconnected minor hemisphere, *Neuropsychologia*, **17** (2), pp. 153–166.
Van Wagenen, W.P. & Herren, R.Y. (1940) Surgical division of commissural pathways in the corpus callosum: Realation to spread of an epileptic attack, *Archives of Neurology & Psychiatry*, **44**, pp. 740–759.
Wolford, G., Miller, M.B. & Gazzaniga, M. (2000) The left hemisphere's role in hypothesis formation, *The Journal of Neuroscience*, **20** (6), pp. 1–4.
Zaidel, D.W. (1994) A view of the world from a split-brain perspective, in Critchley, E. (ed.) *Neurological Boundaries of Reality*, pp. 161–174, London: Farrand Press.
Zaidel, E. (1998) Stereognosis in the chronic split brain: Hemispheric differences, ipsilateral control and sensory integration across the midline, *Neuropsychologia*, **36** (11), pp. 1033–1047.

Martine Nida-Rümelin

The Illusion of Illusionism

Abstract: *A central thesis of Frankish's argument for illusionism is the claim that illusionism is possibly true. This is what the realist about phenomenal consciousness must deny. Frankish's argument for that premise is based on a widely shared understanding of phenomenal consciousness as being a matter of certain events (experiences) instantiating special properties (phenomenal properties). I argue that the illusionist's reasoning is difficult to avoid if one accepts this common account. A positive argument for the thesis that the mere possibility of illusionism can be excluded is developed. It uses a proposal about how the notion of phenomenal consciousness should be taken to pick out the phenomenon it refers to. Given that account it becomes obvious that the illusionist cannot be understood as seriously accepting his own theory. The (supposed) belief in illusionism thus turns out to be an illusion.*

1. Introduction

According to illusionism about phenomenal consciousness, no living being is ever phenomenally conscious. This claim strikes many people as absurd and that reaction is perfectly adequate. There is so much to do in philosophy in search of the truth that one should not lose too much time with absurd theories. But Keith Frankish's defence of illusionism goes wrong in an interesting and illuminating manner. It therefore undoubtedly deserves being seriously discussed.

Frankish identifies the claim that an animal is phenomenally conscious with the claim that it *has experiences which instantiate a certain kind of qualitative properties* which he calls phenomenal

Correspondence:
Martine Nida-Rümelin, University of Fribourg, Switzerland.
Email: martine.nida-ruemelin@unifr.ch

properties. He argues that experiences do not in fact have such properties. But he admits, or so I think we must understand him, that *experiences appear to have such qualitative properties* and that this is why we believe that they do. As will be explained below, these two assumptions lead into illusionism in a quite natural manner. One reason for my claim that Frankish goes wrong in an interesting manner is this: the two claims leading into illusionism are broadly shared or at least not explicitly denied by many realists about phenomenal consciousness. If this diagnosis is adequate then Frankish's argument for illusionism shows that realists about phenomenal consciousness need to abandon central and widely shared presuppositions.

The notion of phenomenal properties so understood is a technical term. Therefore it may seem obvious that their reference must be fixed by theoretical assumptions about them. If the reference of the term 'phenomenal properties' is fixed by such theoretical assumptions and if, as Frankish presupposes, to be phenomenally conscious just *is* to have experiences with phenomenal properties then the question about the existence of phenomenally conscious animals turns into the question about whether experiences have properties satisfying the relevant theoretical assumptions. The realist about phenomenal consciousness should resist this crucial move. To escape the illusionist argument the realist must insist that reference to phenomenal consciousness (more precisely: to being phenomenally conscious where this is a property of individuals) is fixed in a totally different and, importantly, in a *non-theoretical manner*. How reference to phenomenal consciousness is achieved in a non-theoretical manner will be described in Section 3.[1]

I agree with Frankish (on the basis of quite different reasons which are incompatible with his) that experiences do not instantiate

---

[1] A term may be said to be a *theoretical term* if and only if the reference of the term is determined by theoretical assumptions about the referent of the term. Roughly, a theoretical term is not empty if and only if not too many of those assumptions are false. A central point of the present comment may now be put as follows: the terms involved in fixing the reference to phenomenal consciousness are no theoretical terms. Frankish implicitly presupposes that the term 'phenomenal properties' is a theoretical term. This becomes quite clear in Section 1.2, in which he discusses various ways to escape his argument by modifying the notion of phenomenal properties by weakening the theoretical assumptions made about them. He does not seem to consider that the opponent can answer by insisting that reference to phenomenal consciousness must be taken to be established in an altogether non-theoretical manner.

phenomenal properties in the sense here at issue. Experiences do not have qualitative properties of which we are introspectively aware and which constitute what it is like to undergo the experience.[2] But it is one thing to be a non-realist about phenomenal properties in the philosophical sense here at issue and quite another to be a non-realist about phenomenal consciousness. The notion of phenomenal properties when understood as properties of events (experiences) is a confused notion and it has led to various mistakes in contemporary philosophy. The notion of phenomenal consciousness is *not* confused. It is a common mistake to introduce the unproblematic notion of what it is to be phenomenally conscious using the highly problematic notion of phenomenal properties of experiences.[3] Since the target of Frankish's argument is the supposed instantiation of phenomenal properties by experiences, the existence of phenomenal consciousness is no longer under attack if one realizes that the latter must not be identified with the former.

However, in response to Frankish, more must be done. It does not suffice to clearly distinguish between realism about phenomenal consciousness and realism about phenomenal properties of experiences. An argument parallel to the one he develops against realism about phenomenal properties can be formulated against realism about phenomenal consciousness.[4] Frankish argues forcefully that to acknowledge the mere *possibility* that phenomenal properties are never instantiated already provides strong reason to accept that they *actually* are never instantiated. His argument is straightforward: phenomenal properties do not fit into the standard scientific worldview. Therefore, if it is *possible* that phenomenal properties are never instantiated, then everything speaking in favour of the standard scientific worldview speaks in favour of the assumption that they are *actually* never instantiated.

---

[2] I defend this claim in detail in Nida-Rümelin (2016a). The problematic properties (here called 'phenomenal properties') are called qualia in that paper (as they were in earlier versions of Frankish's target article) but this is only a terminological difference which has no impact for the argument developed in that paper.

[3] In Nida-Rümelin (2016a) I develop a detailed argument for this claim.

[4] For simplicity I will be using the term «phenomenal properties» as designating (potential) properties of experiences (of a special sub-class of events). Sometimes the term is used in the literature to refer to properties of experiencing subjects. To clearly distinguish such properties *of conscious individuals* from supposed phenomenal properties *of experiences*, the former properties will be called 'experiential properties' in what follows.

The realist about phenomenal consciousness who admits, as many do (and I am among them), that phenomenal consciousness does not fit into the standard scientific worldview must provide a positive argument for the claim that it is *impossible* that phenomenal consciousness does not exist. Otherwise he or she will have to admit — according to the very same reasoning applied to phenomenal properties by Frankish — that everything speaking in favour of the standard scientific worldview speaks against the existence of phenomenal consciousness.

A proper rejection of illusionism must therefore focus on the illusionist claim that it is *possible* that no one is ever phenomenally conscious. I will proceed here in two steps. In the first part of this comment (Section 2) I argue that the reasons given by Frankish for the assumption that illusionism is possibly true are ill-founded. In the second part (Section 3) I sketch a positive argument against the claim that illusionism about phenomenal consciousness is possibly true.

2. Phenomenal properties and experiential properties

The identification of phenomenal consciousness with the instantiation of phenomenal properties which will be criticized in a moment may be put more precisely in the following way:

PPP (the phenomenal property presupposition):
For a being B to be phenomenally conscious at a given moment m is for it to have experiences at m which instantiate phenomenal properties.

The presupposition PPP plays various roles in Frankish's overall argument. It underlies his thesis that phenomenal consciousness cannot be scientifically accounted for without radical revision. The claim is motivated by reference to supposed features of phenomenal properties (such as their being intrinsic and ineffable). Since I agree for *different* reasons with the claim that phenomenal consciousness poses serious problems for the standard scientific worldview I will not comment here on that role of PPP in the argument for illusionism. Furthermore PPP is presupposed in Frankish's reply to the potential objection that one need not incorporate the problematic features at issue in one's conception of phenomenal properties. Frankish argues that it will be hard or impossible for his opponent to define phenomenal properties in a way which does not turn his or her view into a version of illusionism. This use of PPP can be put aside as well for present purposes. No problem about how the notion of phenomenal properties should be weakened arises for the version of realism about phenomenal consciousness here defended which makes no use of the notion of phenomenal properties in its explication of realism about phenomenal consciousness.

But there is a further and more hidden way in which PPP enters the illusionist argument developed by Frankish. It underlies the crucial idea that *we represent our experiences as having phenomenal properties*. With the general idea that representation involves the possibility of misrepresentation it follows that experiences might not instantiate phenomenal properties.

Frankish's central claim that we represent our experiences as having phenomenal properties is open to various interpretations. First, one might read it as a claim about what we merely believe to be the case as we start thinking about our own experiences. Second, one might read it as the claim that the 'illusion' of experiences as having phenomenal properties occurs as soon as we introspect our experiences by directing our attention towards them. Third, it might be read as the claim that 'we are under the impression' that the experience instantiates phenomenal properties in having the experience itself. According to the third interpretation, the supposed illusion of phenomenality does not require any further step such as introspection or experience-directed thought. I will assume in what follows that the third interpretation is most faithful to what Frankish has in mind. But the argument against his view could be reformulated in a modified version for the other two interpretations as well. On the third interpretation the claim may be stated as follows:

> **RPP:** When a person has an experience, then the experience is represented to her as having phenomenal properties.

I take RPP to be a claim about how things appear to be to the subject concerned. On this understanding RPP states that the experiences we undergo *appear to us to instantiate phenomenal properties* (we are under the impression, in other words, that they instantiate phenomenal properties).

A natural reaction to this claim in the context of an illusionist argument is to wonder: if there is such a thing like being under the impression that experiences have phenomenal properties, isn't it then like something to be under that impression? If so, RPP contradicts the illusionist claim and we can conclude that the illusionist presupposes and actually uses the assumption that phenomenal consciousness exists in his argument against its existence. I will not follow this line of argument here, although I take it to be a promising one. Let us ask instead: is RPP true?

In order to judge RPP we must first ask what experiences are and how if at all one can make sense of the claim that they instantiate phenomenal properties. Experiences are special kinds of events. Your blue experience as you look at the sea is an example. What does it consist in? The obvious answer is: it consists in your instantiation of a certain property. You are visually presented with blueness. The experience at issue consists in the fact that *you* have *that property* at the moment at issue. Not all instantiations of properties by people or animals are experiences. Let us call the relevant subclass of properties *experiential properties*.

Phenomenal properties are standardly described as *those properties of experiences in virtue of which it is like something for the subject concerned to have the experience.* So we must ask: what property of your blue experience is it in virtue of that it is something it is like for you to 'have it' (it would be better to say: 'to be involved in it' or 'to undergo it')? Well, the answer seems obvious: it is like something for you to undergo a blue experience in virtue of the fact that blue experiences are instantiations of special properties, namely of *experiential properties*, properties which are such that having them is like something for the individual who has them. *The whole talk of phenomenal properties of experiences can be given an understandable sense only by reducing it to talk about the properties instantiated by people or other experiencing subjects.* To say that an experience of a given phenomenal kind instantiates a specific phenomenal property (or, as it is often said, a specific quale) can only mean this: experiences of that kind are instantiations of a specific experiential property.

Blue experiences are supposed to instantiate a specific phenomenal property (sometimes referred to talking of 'a blueish quale'). But that 'phenomenal property' supposedly shared by blue experiences can only be, as just argued, *the property of the experience to be such that it consists in the instantiation of the specific experiential property 'being visually presented with blueness' by some experiencing being.* Let us call this property of blue experience the property PEB. PEB is not itself a qualitative property in any clear sense. It must not be confused with the blueness itself which is presented to the experiencing subject. That hue is properly described as qualitative but it is surely not identical to PEB. PEB, furthermore, must of course not be confused with the property of the subject concerned who is involved in the event, with the property of being presented with blueness. *That* property is qualitative in an understandable sense but is not the property PEB.

The argument here briefly presented and elaborated in detail in Nida-Rümelin (2016a) shows that experiences do not have qualitative properties and that the notion of phenomenal properties is fundamentally confused. It shows, furthermore, that the discussion about the status of phenomenal consciousness should not be couched in terms of phenomenal properties but rather in terms of experiential properties. If this is the view to adopt, what follows for the claim RPP according to which experiences appear to have phenomenal properties?

The obvious consequence is that RPP is based on confusion as well. First, since RPP is supposed to describe the states in which we are under the supposed illusion of phenomenality, RPP should adequately describe the way in which it *appears to us that we are phenomenally conscious*. But if to be phenomenally conscious cannot be captured by saying that it is to have experiences with phenomenal properties, the supposed apparent instantiation of phenomenal properties cannot be the way in which we *appear* to be phenomenally conscious. Second, RPP portrays us *as being presented with the experience in which we are involved. But the experience in which we are involved is not itself an object of our experience.* Third, there is no reason to assume that the event in which we are involved when we instantiate an experiential property appears to us to have qualitative properties. The only reason for supposing so is the mistaken assumption that phenomenal consciousness, if it exists, is the instantiation of phenomenal properties by those events. Once this assumption is seriously abandoned, RPP will be seen as totally unmotivated.[5]

It follows that Frankish's motivation for the claim that it is possible that phenomenal consciousness does not exist is unfounded since it is based on a mistaken view of what phenomenal consciousness is supposed to be (the claim PPP) and on a related mistaken view of what it is to be 'under the impression' of being phenomenally conscious (the claim RPP).

According to the alternative here proposed the claim that consciousness exists is to say nothing less and nothing more than this: some individuals sometimes instantiate experiential properties. The

---

[5] For a detailed refutation of PPP and RPP see Nida-Rümelin (2016a), in particular Sections 9–12.

illusionist might hope at this point that he can simply repeat his argument saying something like this: our belief in the instantiation of experiential properties is justified by our apparent awareness of having such properties. To be (apparently) aware of having experiential properties is to represent oneself as having those properties. Representations can misrepresent. Therefore, it might be that no one ever has any experiential property.

An indirect way to respond to this is to give a positive argument for the claim that the notion of experiential properties can and should be introduced in a way which excludes that no one ever has experiential properties. I will sketch such an argument in the next section. A direct way to respond is to further develop the notion of experiential properties. Their nature must be explained in a way which blocks the argument from possible misrepresentation sketched above. How can that be done?

My reply can be briefly put as follows: to have an experiential property *is* to be aware of having it. This is the remarkable feature of experiential properties. In virtue of this remarkable feature of experiential properties, we are aware of having experiential properties in a way which is not representational in a sense that allows for misrepresentation. If to have a property P and to be aware of having P is one and the same thing, then the awareness of having P cannot possibly 'misrepresent' oneself as having P. On that view, being aware of having an experiential property by having that experiential property does not involve any further step (no reflection, no introspection, no conceptualization) and therefore leaves no room for any kind of illusion. To be aware, in the relevant sense, of having experiential properties is to be aware of being phenomenally conscious. Since awareness of having an experiential property does not allow for any illusion, it follows that we are aware of being phenomenally conscious in a way which does not allow for illusion either.

Let me briefly state how the view just sketched can be motivated and based on an account of experiential properties. Experiential properties are such that, by their nature, the instantiation of such a property by a subject at a given moment m partially constitutes what it is like for the subject to live through moment m. Now there is an obvious and clear sense in which any subject is aware of what it is like for it to live through moment m just by living through that moment. One may say that the subject is immediately aware — without reflection and without conceptualization — of every specific aspect of what it is like for it to live through moment m while it is living through moment m. To admit this is already to admit what needs to be seen: given the nature of experiential properties, to be aware of such an aspect *is* to be aware of having the corresponding experiential property. And to be aware of having a given experiential property is to be aware of being phenomenally conscious.[6]

3. Fixing reference to phenomenal consciousness

To argue for the claim that illusionism *cannot* be true requires an account of how reference to phenomenal consciousness is established. According to the view proposed here no theoretical assumptions about what phenomenal consciousness is play any crucial role in fixing the reference of that term or, to put it more precisely and in the terminology above-motivated: the extension of the notion of experiential properties is not determined by any theoretical assumptions about what experiential properties are. Rather, the term must be taken to be introduced by typical examples and a preliminary, pre-theoretical understanding of what those examples have in common.

The notion of experiential properties can and should be introduced in two steps. In *step 1* a list of paradigmatic examples of such properties is established (the list will include examples like being visually presented with blueness, suffering an intense pain, being sad, experiencing oneself as active, imagining a pink elephant flying above a blue lake, etc.). In *step 2* one aims at establishing shared reference to a feature those examples have in common. One may use metaphors in that step and preliminary descriptions of the common feature one has in mind. It is not important that the descriptions used survive later theoretical progress. Everything that may help to establish shared reference is good enough. Here is one such description of the common feature in question: for all those properties, it is like something for the person or animal to have that property and having the property at least partially consists in what it is like to have it. The description appears to be very successful for many people in making it clear what common feature it is supposed to pick out. The description should, however, not be read as an attempted reduction of the notion of experiential properties to an already understood notion of 'what-it's-like-ness'. It is just a way to 'point to' a common feature typical for

---

[6] For an elaboration of this view see Nida-Rümelin (2016b, Section 6; and in preparation, Section 6).

the examples given in the first step; there are many other ways to do so.

According to the here proposed view, to say that phenomenal consciousness does not exist is to say that no one ever has experiential properties. This claim can only be true if either (a) the way the notion of experiential properties is introduced by the two steps just mentioned fails or (b) although these two steps successfully pick out a set of properties, no member of that set is ever instantiated. Is it possible that one of these conditions is satisfied?

The examples on the list in step 1 are, by stipulation, experiential properties. If at least one of these properties is ever instantiated, illusionism about phenomenal consciousness is false. I am quite certain that Frankish does not wish to deny that people sometimes have colour experiences, that they are sometimes sad, and that many animals sometimes suffer pain. He only wishes to deny that undergoing these experiences involves qualia. If I am right in this assumption then Frankish does not in fact deny the existence of phenomenal consciousness. He is under the illusion of denying it and the illusion is based on a distorted view of what phenomenal consciousness is supposed to be.

Since the notion of experiential properties is introduced by paradigmatic examples, option (b) above can only be realized if step 1 and step 2 are successful and yet the properties referred to in step 1 are never instantiated. This position is difficult to defend since it is hard to explain how step 2 could be successful if we never have experiential properties. In any case, to claim that illusionism is true because (b) is the case is to endorse the radical position that no one ever has colour experiences, emotional experiences, etc. This is not, as far as I can see, the position endorsed by Frankish.

The only remaining option is then that the introduction of the class of experiential properties fails in the second step while step 1 is successful. A failure in step 2 would mean that we do not succeed in establishing shared reference to a common feature of experiential properties. If this is true, then experiential properties are not a homogeneous class; we only seem to be able to develop a positive understanding of their common feature but really in developing that understanding we do not succeed in gaining cognitive access to any substantial commonality among them. This is perhaps the closest one can get to an illusionist view if one accepts the story told here about how reference to phenomenal consciousness is established. But note that it is not a position which deserves being called illusionism about phenomenal consciousness. It is one thing to say that experiential properties do not share any substantial features we understand in the second step mentioned above and quite another thing to say that experiential properties are never instantiated. So to claim the possibility that there is no phenomenal consciousness is to claim that no one ever has any of the paradigmatic examples of experiential properties.

In order to save his claim that some relevant illusion is involved in the way we are aware of being phenomenally conscious, the illusionist might be tempted to respond that we introspectively seem to have experiential properties which are of a nature that makes them unaccountable in physical terms while in fact they *are* accountable in physical terms. On that view experiential properties present themselves to us upon reflection in a way which induces mistaken beliefs about their nature. But this reply does not serve the illusionist's purposes. It is one thing to say that, introspectively, experiential properties appear to be of a nature which is not their real nature, and quite another to say that experiential properties are not instantiated at all. The illusionist must defend the latter claim and must therefore choose between the two unattractive options discussed above.

4. The illusion of illusionism

I announced an argument for the claim that illusionism is not possibly true. To be more precise, one must say that the result obtained is a conditional one: if anyone has ever one of the properties that one may plausibly put on the list in step 1, then illusionism is false. We thus can exclude that illusionism is true if we accept a very weak assumption: some people sometimes feel pain, are happy, sad, or enjoy the smell of basil, etc. In fact it suffices to concede that there is at least one such property which is sometimes instantiated. It is crazy to deny this. Therefore illusionism is an absurd view and the initial reaction mentioned at the beginning is perfectly adequate.

Illusionists, I take it, do not deny the antecedent of the conditional and therefore do not in fact deny the existence of phenomenal consciousness. If they think they do, this is a cognitive illusion. How is it possible that a sophisticated philosopher and a thoughtful person like Keith Frankish and many others do not realize that they are under an illusion when they take themselves to believe in the non-existence of phenomenal consciousness? Despite the efforts I have made here to analyse the mistakes involved in illusionism which partially explain how one can seriously believe in the truth of illusionism, this fact remains nonetheless a mystery to me. It seems so easy to understand what those who talk about phenomenal consciousness — even if they couch it in strange terminology involving the notion of qualitative properties of experiences — really have in mind. Understanding what they really have in mind should be sufficient, I would say, to realize that its existence cannot be reasonably denied.

Acknowledgments

I would like to thank David Chalmers and an anonymous referee for their criticism on an earlier version of this paper which helped me a great deal in arriving at the present and rectified version of the anti-illusionist argument.

References

Nida-Rümelin, M. (2016a) The experience property framework — a misleading paradigm, in Ben-Yami, H. (ed.) *Trends in Philosophy of Language and Mind*, special issue of *Synthese*, in print.

Nida-Rümelin, M. (2016b) Self-awareness, in Farrell, J., *et al.* (eds.) *Consciousness and Inner Awareness*, special issue of *Review of Philosophy and Psychology*, in print.

Nida-Rümelin, M. (in preparation) Experiencing subjects and so-called mineness, in Garcia Carpinteiro, M. & Guillot, M. (eds.) *The Sense of Mineness*, Oxford: Oxford University Press.

Derk Pereboom

Illusionism and Anti-Functionalism about Phenomenal Consciousness

Abstract: *The role of a functionalist account of phenomenal properties in Keith Frankish's illusionist position results in two issues for his view. The first concerns the ontological status of illusions of phenomenality. Illusionists are committed to their existence, and these illusions would appear to have phenomenal features. Frankish argues that functionalism about (quasi-)phenomenal properties yields a response, but I contend that it doesn't, and that instead the illusionist's basic account of phenomenal properties must be reapplied to the illusions themselves. The second concern is that phenomenal properties would seem to be intrinsic properties of experience, but functionalism has them consist solely in relations. The nonfunctionalist option can recapture the sense that these properties are intrinsic. It can also preserve the intuition that they are causal powers in a robust sense, and thereby, perhaps surprisingly, provide a stronger response to the knowledge and conceivability arguments.*

1. Introduction

I generally agree with Keith Frankish's very fine exposition and defence of the illusionist option about phenomenal consciousness (for perhaps the earliest version see Dennett, 1988; *cf.* Churchland, 1981; Humphrey, 1992), and with his account of its theoretical advantages

Correspondence:
Derk Pereboom, Cornell University, Ithaca, New York, USA.
Email: dp346@cornell.edu

over radical and conservative realism. As Frankish notes, there are strong exclusionary pressures against dualist radical realism. Such a radical realism may deny the causal closure of the physical, but not without cost. And I agree that conservative realism, in particular the phenomenal concept strategy version, is subject to the kinds of instability concerns that Frankish sets out, although the issues are complex and undoubtedly more remains to be said. There are two issues on which I want to comment, and both concern the functionalist interpretation of the illusionist option. That is, both concern whether the strongest sort of illusionism opts for the view that phenomenal properties or their quasi-phenomenal correlates, to use Frankish's term, are best cast as consisting exclusively of causal relations to sensory inputs, behavioural outputs, and other mental states. I generally favour a non-functionalist account of the mental for the reason that it accords best with a non-Humean account of causal powers generally, including mental causal powers (Pereboom, 1991; 2002; 2011).

Functionalism has two roles in Frankish's account that I will challenge. The first concerns the ontological status of illusions of phenomenality themselves. Illusionists are committed to their existence, and these illusions would appear to have phenomenal features. The second is that phenomenal properties would seem to be intrinsic properties of experience, but functionalism about these properties would have them consist solely in relations, that is, extrinsic features. The non-functionalist option can preserve the sense that the reality accessed is in fact an intrinsic feature of experience, and this, I argue, has a number of advantages.

2. Do illusions have phenomenal properties?

So first, illusionists agree that the what-it's-like features of sensory states, what I call the qualitative natures of phenomenal properties, are illusory in that they don't exist. But what of the illusion itself? Experiencing an illusion requires that the illusions themselves, which are kinds of introspective representations, will exist. As Frankish himself remarks, 'Forming the theoretical belief that phenomenal properties are illusory does not change one's introspective representations' (p. 18). However, these illusions themselves will have phenomenal properties, or so it would seem. Illusions of phenomenal properties would appear to differ in a key respect from other sorts of illusions, for instance from the illusions of psychokinesis that Frankish adduces. Illusions of psychokinesis will not have psychokinetic properties, while illusions of phenomenal properties would seem to have phenomenal properties. Accordingly, the objector may claim that there will be something it's like to have an illusion of phenomenal greenness, and it's the same as what it's like have a sensation of green. If this is right, then the illusionist's strategy won't get us very far (see Pereboom, 1994, pp. 582–4; Alter, 1995; and Chalmers, 1996, p. 142, for versions of this objection).

In setting out his account of the concern, Frankish says: 'The objector may reply that in order to create the illusion of a greenish experience, the introspective representation would have to apply a greenish mode of presentation, which would itself have an introspectable greenish feel.' In his response he contends that 'illusionists will simply deny this, arguing that the content of introspective representations is determined by non-phenomenal, causal or functional factors' (pp. 32–3). However, this seems insufficient. The objector may even agree that the content of the illusory experience is functional, but point out that the illusion nevertheless exists, and argue there is something it's like to have it.

Frankish points out in a footnote (p. 33, n. 13) that I set out an answer to this objection that reapplies the illusionist's hypothesis to such modes of presentation. I think that this answer is essential to the strength of the illusionist's case. The proposed alternative, claiming functionalist content for the illusory representation, appears not to address the objection effectively.

Let me explain my response in some detail, and I'll begin with a summary of my way of setting out illusionism. The core concern about physicalism expressed in the both Frank Jackson's (1980; 1986) knowledge argument and David Chalmers' (1996; 2002) conceivability argument begins with the claim that phenomenal states have phenomenal, what-it's-like type, properties, and concerning them it seems plausible that:

(i) Introspective modes of presentation represent a phenomenal property as having a specific qualitative nature, and the qualitative nature that an introspective mode of presentation represents a phenomenal property as having is not among the features that any physical mode of presentation would represent it as having.

Moreover, it is also seems plausible that:

(ii) The introspective mode of presentation *accurately represents* the qualitative nature of the phenomenal property. That is, the introspective mode of presentation represents the phenomenal property as having a specific qualitative nature, and the attribution of this nature to the phenomenal property is correct.

One way to characterize the qualitative natures that introspective modes of presentation represent phenomenal properties as having is by way of resemblance to modes of presentation (*cf.* Locke, 1689/1975): Mary's introspective representation of her phenomenal-red sensation presents that sensation in a characteristic way, and it is plausible that a qualitative nature that resembles this mode of presentation is accurately attributed to the sensation's phenomenal property. Alternatively specified, this qualitative nature is just as the introspective mode of presentation represents it to be.

Given these claims about what is least initially plausible, the proponent of the knowledge argument can explain its force in the following way. Mary, upon leaving her black-and-white room and seeing the ripe tomato, comes to have the belief:

(A) Seeing red has R,

where 'R' is the phenomenal concept that directly refers to phenomenal redness, that is, to phenomenal property R. We'll assume that the qualitative nature of phenomenal redness is accurately represented introspectively by Mary's introspective phenomenal mode of presentation. We'll also assumes that, if physicalism is true, then every truth about the qualitative nature that an introspective mode of presentation accurately represents a phenomenal property as having will be derivable from a proposition detailing only features that physical modes of presentation represent the world as having (an assumption I don't challenge). However, (A) is not derivable from this proposition. So not every truth about the qualitative nature that introspective modes of presentation accurately represent phenomenal properties as having is so derivable. Thus physicalism about phenomenal properties is false.

This version of the knowledge argument can be challenged by questioning one or both of the claims about what is intuitive. The illusionist takes issue with (ii), the claim about the accuracy of introspective phenomenal representation. Supposing the truth of (i), it is an open possibility that introspective representation is inaccurate in the sense that it represents phenomenal properties as having qualitative natures they in fact lack; that is, it is an open possibility that the *qualitative inaccuracy hypothesis* is true. Upon seeing the red tomato, Mary introspectively represents the qualitative nature of phenomenal redness in the characteristic way, but it is an open possibility that she is representing phenomenal redness as having a qualitative nature that it actually does not have. Here, then, is the illusion.

As Frankish notes, the seriousness of this open possibility can be supported by an analogy with our visual representation of colours (Pereboom, 1994; 2011). Our visual system represents colours as having certain specific qualitative natures, and it is an open possibility, widely regarded as actual, that colours actually lack these natures. It's open that introspection of phenomenal properties involves representational systems relevantly similar to those involved in visual colour perception, and this would explain how the qualitative inaccuracy hypothesis about the phenomenal could be true. It's plausible that part of what explains qualitative inaccuracy about colour representation is that it's causal, and this allows for a discrepancy between the real nature of colours and how we represent their qualitative natures. It might well be that introspective representation of phenomenal colour is also causal, and that this gives rise to a similar discrepancy. A terminological and perhaps substantive point: Frankish opts for the view that the what-it's-like qualitative nature is what phenomenality is. So what remains given inaccuracy and the resulting illusionism is a quasi-phenomenal correlate or a quasi-phenomenal cause of the introspective representation. But, as Frankish notes, abandoning the view that physical objects have properties that resemble our visual representations of colour doesn't motivate that the colour correlate that remains is not really colour but only 'quasi-colour'. Rather, colour turns out to differ from what we might have naturally assumed it to be. Analogously, the illusionist can argue that phenomenal properties are preserved given her view, but they turn out to differ from what we ordinarily assume them to be. But even if this is right, Frankish's terminology is valuable for clarity of exposition, and I will use it in what follows.

Supposing that this illusionist open possibility is in fact realized, how should we characterize what happens when Mary leaves the room and sees the red tomato? She now comes to have a belief of the form:

(A) Seeing red has R.

Consider first the proposal that the concept 'R' in this belief refers to a property with the qualitative nature accurately represented by that characteristic mode of presentation. Supposing our open possibility is actual, seeing red lacks this qualitative nature, and as a result belief (A) will be false. Then coming to believe (A) wouldn't amount to Mary's learning something about the qualitative nature of phenomenal redness, that is, to acquiring a new true belief about it. Alternatively, consider the proposal that 'R' refers to a quasi-phenomenal property, which is introspectively misrepresented and lacks such a qualitative nature, and is uncontroversially entirely physical. We can suppose that this belief is then true, but given this supposition Mary already had this belief when in her room, or was able to derive it from the true beliefs then had, she also does not acquire a new true belief. So either Mary's belief (A) about seeing red wouldn't be true or wouldn't be new.

It is at this point that we encounter the objection that illusions of phenomenality themselves have phenomenal properties. Thus it might be objected that the account as explained up to this point merely shifts the problem for a physicalist explanation from accounting for phenomenal properties to accounting for the introspective phenomenal modes of presentation that characterize illusions of phenomenal properties. Given the qualitative inaccuracy hypothesis, the object of Mary's sensation indeed lacks the qualitative nature that would be accurately represented by her introspective representation, and thus she cannot learn that this object has such a qualitative nature. However, she nevertheless seems to learn something about how the illusion's mode of presentation — let's call it 'MPR' — represents what it does. And it's at least initially plausible that Mary cannot derive the corresponding truth about how MPR represents from her microphysical base.

Here is my response to this objection (2011, pp. 27–8; *cf.* Pereboom, 1994). If the qualitative inaccuracy hypothesis applies to introspective representations of (quasi-)phenomenal states, it makes sense to suggest that it also applies to introspective representations of phenomenal modes of presentation. It would then also be an open possibility that Mary introspectively represents MPR as having a qualitative nature it really lacks. Then, despite how MPR is introspectively represented, she would be able to derive every truth about its real nature from her microphysical base even while in her black-and-white room.

When I've presented this response to various audiences, I almost always encounter the objection that it generates an unwelcome infinite regress of some kind. For example, one might suspect that, given this response, when I represent an introspective representation, I do so by way of a mode of presentation, whereupon I would represent this mode of presentation by another introspective mode of presentation, and I would represent that further mode of presentation by a yet further introspective mode of presentation, *ad infinitum*. Thus when I represent an introspective representation, I would in fact have an infinite series of introspective representations, and this is absurd.

The plausibility of my response to the objection that illusions of phenomenality are themselves phenomenal requires that, when Mary introspectively represents her sensation of red, she also introspectively represents the introspective mode of presentation MPR of that sensation, while no actual infinite regress of introspective representations is generated. A key point is that a mode of presentation can function as the way a subject represents something without that subject also representing the mode of presentation itself. She might, in addition, represent this mode of presentation, but this would be a distinct representation that is not necessitated by her initial representation. If she did represent the mode of presentation, it could be by means of a higher-order introspective mode of presentation. However, it would not be necessitated that she also represents this higher-order mode of presentation.

In addition, it's credible that, when someone introspectively represents a sensation of red by way of MPR, she will normally, although not necessarily, also represent MPR by a higher-order mode of presentation, but only in unusual cases will she introspectively represent that higher-order mode of presentation, or modes of presentation at yet higher orders. Moreover, as Nico Silins suggested to me (in conversation), it is implausible that beyond some fairly low level of iteration mental states would be introspectively represented by way of *phenomenal* modes of presentation. At some level of iteration, I am able only to form a belief, without any distinctive phenomenology, that I am representing a mental state. Such a belief, because it is not phenomenal, would fail to lend support to the knowledge argument.

Thus by reapplying the illusionist's basic strategy to the illusion itself, a key objection to illusionism can be answered, and given this reply, no actual infinite regress is generated.

3. The advantages of retaining intrinsic quasi-phenomenal properties

My second concern is that phenomenal properties would seem to be intrinsic properties of experience, but illusionism supplemented by functionalism about phenomenal or quasi-phenomenal properties would have them consist solely in relations to causes and effects. The non-functionalist option can respect the intuition that phenomenal properties are in fact intrinsic features of experience, and also honour the sense that they are non-Humean causal powers. In addition, it provides a further response to the knowledge and conceivability arguments as set out by Jackson and Chalmers. True, illusionism already yields a response to these anti-materialist arguments without invoking non-functionalism. But the additional response bolsters the case against these arguments, and my sense is that preserving intrinsicness makes the overall view more plausible.

Let me begin with the response non-functionalism provides to the anti-materialist arguments. Chalmers' (2002; *cf.* 1996) exposition of his conceivability argument allows us to see this advantage most clearly: let 'P' be a statement that details the complete physical truth about the actual world; 'T' a 'that's all there is' statement, specifying that P describes a minimal P-world; and 'Q' an arbitrarily selected actual phenomenal truth. Statement S is ideally conceivable when it is conceivable on ideal rational reflection, and *prima facie* conceivable when conceivable on initial reflection. S is primarily possible just in case it is true in some world considered as actual. S is secondarily possible just in case S is true in some world considered as counterfactual. Correlatively, S is primarily conceivable just in case S can be conceived as true in some world considered as actual, or alternatively, since considering-as-actual is an *a priori* matter, S is primarily conceivable just in case the subject can't rule out S *a priori*. Given these preliminaries, here is the argument:

(1) 'PT and ~Q' is ideally primarily conceivable.
(2) If 'PT and ~Q' is ideally primarily conceivable, then 'PT and ~Q' is primarily possible.
(3) If 'PT and ~Q' is primarily possible, then 'PT and ~Q' is secondarily possible, or Russellian monism is true.
(4) If 'PT and ~Q' is secondarily possible, materialism is false.
(5) Materialism is false or Russellian monism is true.

Russellian monism is any view that combines (1) *categorical ignorance*, the claim that physics leaves us ignorant of certain categorical bases of physical dispositional properties, with (2) *consciousness- or experience-relevance*, the proposal that these categorical properties have a significant role in explaining consciousness or experience.

The Russellian monist escape is key to the advantage I have in mind. Here is how Chalmers conceives this escape. While primary/*a priori* conceiving of P fixes all of the dispositional and relational properties designated by physical concepts, it does not determine categorical and intrinsic properties that underlie and explain them. So it's open that one can primarily conceive 'PT and ~Q' only because one is conceiving just dispositional and relational properties on the physical side, and it is an open possibility that if one were to replace 'P' with a more complete 'P*' that includes concepts that allow for direct representation of the currently unknown or incompletely understood intrinsic properties in their categorical basis, 'P*T and ~Q' would not be primarily conceivable. 'Q' would then be *a priori* derivable from 'P*T', and P* would explain phenomenal consciousness (Chalmers, 2002; 2003; Stoljar, 2006; Pereboom, 2011; Alter and Nagasawa, 2012).

The following is a crucial issue for the success of the argument. Suppose we agree that:

(0) 'PT and ~Q' is *prima facie* primarily conceivable.

Can we secure:

(1) 'PT and ~Q' is ideally primarily conceivable?

Without (1), we can't advance to primary possibility. But we're not ideal conceivers, so there's a problem. But according to Chalmers, what licenses the move from (0) to (1) is the thesis that from truths solely about structure only truths solely about structure can be derived — the *from-structure-only-structure* thesis (Chalmers, 2003; Stoljar, 2006; Pereboom, 2011; Alter, 2016). Structural properties are, on first pass, relational, i.e. extrinsic. (Dynamical properties, also invoked in this context, are alterations in structural properties over time, and they are also relational, and we can think of them as a kind of structural property.) On a first, rough interpretation of this thesis, it tells us that only truths solely about relational properties can be derived from truths solely about relational properties. On the assumption that the physical is purely structural and thus relational, and that phenomenal properties are intrinsic properties of experience and thus not relational, the relevant truths about phenomenal properties won't be derivable from physical truths, no matter how much we come to know about the physical. Even in the ideal case, the relevant phenomenal truths won't be derivable from physical truths.

But this clearly won't work, since much of our physical knowledge is in fact about intrinsic and thus non-relational properties. For example, a ball's shape and rigidity are intrinsic properties of the ball, and physics does not leave us ignorant of them. At this point I've propose the following refinement (2011, pp. 92–4). The categorical bases of which physics leaves us ignorant (on Chalmers' Russellian monist conception) are intrinsic properties of a fundamental kind — *absolutely intrinsic properties*.

> P is an *absolutely intrinsic* property of X just in case P is an intrinsic property of X, and P is not fully grounded in extrinsic properties of parts of X.

By contrast

> P is a *(merely) relatively intrinsic* property of X just in case P is an intrinsic property of X, and P is fully grounded in extrinsic properties of parts of X.

(For X to ground Y, X must necessitate X and Y must exist in virtue of Y.) The notion of an absolutely intrinsic property facilitates a more plausible and precise characterization of the from-structure-only-structure thesis. Consider an objection that Daniel Stoljar raises against this principle:

> The simplest way to see that the from-structure-only-structure thesis is false is to note that one can derive the instantiation of an intrinsic property from a relational one just by shifting what thing you are talking about. For example, being a husband is a relational property of Jack Spratt, and being a wife is a relational property of his wife. But being married is an intrinsic property of the pair (or the sum) of Jack Spratt and his wife. To take a different example, it seems plausible to say that I have the property of having a hand intrinsically, but my having this property obviously follows from a relation between my hand and the rest of my body, and that the truth concerning this is a relational truth. (Stoljar, 2006, p. 152)

Torin Alter (2009) agrees that Stoljar has a point: if objects x and y compose object z, then it is possible to derive intrinsic properties of z from relational properties of x and y. But Alter argues that this objection undermines the from-structure-only-structure thesis only if non-structural properties are identified with intrinsic properties, and in his view that identification is mistaken, for 'the property *being married* is purely structural/dynamic despite being intrinsic to the Spratts. Any structural/dynamical duplicate of the actual world contains a corresponding married pair' (*ibid.*). Alter contends that such examples indicate not that we should reject the from-structure-only-structure thesis, but rather that we should not identify non-structural properties with intrinsic properties.

How might *being a married pair* be a structural property of the Spratts? The distinction between relatively and absolutely intrinsic properties supplies a diagnosis. Although *being a married pair* is an intrinsic property of the Spratts, it is fully grounded in Jack's extrinsic property of *being married to Jill* and Jill's extrinsic property of *being married to Jack*. Consequently, *being a married pair* is a relatively intrinsic property and not an absolutely intrinsic property of the Spratts. I propose, then, that all non-structural properties are absolutely intrinsic properties (or have absolutely intrinsic components). Stoljar's counter-example would then fail against the from-structure-only-structure thesis, and, more generally, this principle would be in the clear (Pereboom, 2011, pp. 110–14).

But notice that, given that the from-structure-only-structure thesis is what facilitates the transition from

(0) PT and ~Q' is *prima facie* primarily conceivable

to

(1) PT and ~Q' is ideally primarily conceivable,

a non-structural conception of the physical would block it. Chalmers allows Russellian monist conceptions of the physical, which would function as an escape from the argument, but the examples he adduces are restricted to versions of pan- or micropsychism (e.g. Strawson, 2003) or involve currently unconceived intrinsic properties that are similar enough to paradigmatic physical properties to count as physical themselves. Illusionists might well resist any such proposals as being too mysterious. But there is a non-mysterious proposal for absolutely intrinsic physical properties to which illusionism *per se* wouldn't be averse. It crucially features a non-Humean, realist view of causal powers.

Causal powers are often cast as dispositional properties. But dispositions can be characterized as mere tendencies, to be explicated solely in terms of subjunctive conditionals. For a model, take the tendency of a ball to roll when pushed to be a dispositional property. The ball has a disposition such that if it were located on a plane surface and pushed, it would roll. It's often argued that dispositions conceived in this way require categorical bases to explain the dispositional tendencies. The ball's shape and rigidity qualify as such categorical bases. Typically, those categorical bases would be intrinsic properties of the entity that has them. On the non-Humean view, causal powers are identified with the intrinsic categorical bases, and not with the mere tendencies.

Jonathan Jacobs (2011; *cf.* John Heil, 2003) explains this non-Humean notion of a causal power as that of an intrinsic property, not itself explicable in terms of subjunctive conditionals, that serves as a truthmaker for such subjunctive conditionals. The ball's shape and rigidity, both intrinsic properties, are truthmakers for why it is that, if the ball were located on a plane surface and pushed, it would roll. On Jacobs' view, this grounding in such a truthmaker explicates how it is that the categorical basis is a causal power. One might call such a categorical basis, following Sydney Shoemaker (in conversation), an *intrinsic aptness* to produce certain effects.

In Chalmers' conceivability argument we're thinking of P as the complete microphysical truth. Let's agree that the most fundamental forces that current physics invokes — gravitation, electromagnetism, and the strong and weak nuclear forces — are described by physics in exclusively dispositional/relational terms, and that those dispositional descriptions are cast in terms of subjunctive conditionals such as: if an electron were in the vicinity of another electron, it would repel that electron. On Jacobs' account of causal powers, they are intrinsic properties that serve as truthmakers for such subjunctive conditionals, and we can call them intrinsic physical aptnesses to produce specific effects. We can now add that these truthmakers would consist at least in part of *absolutely* intrinsic properties. And with such non-mysterious non-structural properties in the base (P*), the from-structure-only-structure principle won't license the move from *prima facie* to ideal conceivability of 'P*T and ~Q'.

Thus if the illusionist were to accept that phenomenal or quasi-phenomenal properties consist at least in part of absolutely intrinsic aptnesses, she'd have a stronger response to the conceivability argument (and to the knowledge argument as well, since it implicitly requires the claim that Mary, as ideal reasoner, can't derive 'Q' from 'PT', and we, as non-ideal reasoners, would need the from-structure-only-structure thesis in order to ascertain this claim). But this is inconsistent with functionalism about such properties, for then they would consist solely in relations. The illusionist would also have a response to the claim that phenomenal properties are intrinsic properties of experience, which she can now accept. Finally, it's intuitive that phenomenal properties are causal powers. Phenomenal pain and phenomenal pleasure would seem to be powers to cause aversive and attractive responses. It's arguably intuitive, at least for many, that phenomenal properties are causal powers on the non-Humean model. If the reality of phenomenal pain consisted in a functional property, it would consist solely in relations to causes and effects, and then it could not be a causal power on the non-Humean model. To the extent that it is intuitive that pain is such a causal power, the non-functional account of phenomenal or quasi-phenomenal properties has the advantage.

References

Alter, T. (1995) Mary's new perspective, *Australasian Journal of Philosophy*, **73**, pp. 582–584.

Alter, T. (2009) Does the ignorance hypothesis undermine the conceivability and knowledge arguments?, *Philosophy and Phenomenological Research*, **79**, pp. 756–765.

Alter, T. (2016) The structure and dynamics argument against materialism, *Noûs*, **50** (4), pp. 794–815.

Alter, T. & Nagasawa, Y. (2012) What is Russellian monism?, *Journal of Consciousness Studies*, **19** (9–10), pp. 67–95.

Chalmers, D. (1996) *The Conscious Mind*, New York: Oxford University Press.

Chalmers, D.J. (2002) Does conceivability entail possibility?, in Gendler, T. & Hawthorne, J. (eds.) *Conceivability and Possibility*, pp. 145–200, Oxford: Oxford University Press.

Chalmers, D.J. (2003) Consciousness and its place in nature, in Stich, S. & Warfield, T. (eds.) *Guide to the Philosophy of Mind*, Oxford: Blackwell.

Churchland, P.M. (1981) Eliminative materialism and the propositional attitudes, *Journal of Philosophy*, **78**, pp. 67–90.

Dennett, D. (1988) Quining qualia, in Marcel, A. & Bisiach, E. (eds.) *Consciousness in Modern Science*, pp. 42–97, New York: Oxford University Press.

Frankish, K. (this issue) Illusionism as a theory of consciousness, *Journal of Consciousness Studies*, **23** (11–12).

Heil, J. (2003) Dispositions, *Synthese*, **144**, pp. 343–356.

Humphrey, N. (1992) *A History of the Mind: Evolution and the Birth of Consciousness*, New York: Simon and Schuster.

Jackson, F. (1980) Epiphenomenal qualia, *Philosophical Quarterly*, **32**, pp. 127–136.

Jackson, F. (1986) What Mary didn't know, *Journal of Philosophy*, **83**, pp. 291–295.

Jacobs, J. (2011) Powerful qualities, not pure powers, *The Monist*, **94**, pp. 81–102.

Locke, J. (1689/1975) *An Essay Concerning Human Understanding*, Oxford: Oxford University Press.
Pereboom, D. (1991) Why a scientific realist cannot be a functionalist, *Synthese*, **88**, pp. 341–358.
Pereboom, D. (1994) Bats, brain scientists, and the limitations of introspection, *Philosophy and Phenomenological Research*, **54**, pp. 315–329.
Pereboom, D. (2002) Robust nonreductive materialism, *Journal of Philosophy*, **99**, pp. 499–531.
Pereboom, D. (2011) *Consciousness and the Prospects of Physicalism*, New York: Oxford University Press.
Stoljar, D. (2006) *Ignorance and Imagination*, New York: Oxford University Press.
Strawson, G. (2003) Real materialism, in Antony, L. & Hornstein, N. (eds.) *Chomsky and His Critics*, Oxford: Blackwell.

Jesse Prinz

Against Illusionism

Abstract: *Illusionism is the view that phenomenal qualities are an illusion. It contrasts with both dualist theories and reductive realist theories, which identify phenomenal qualities with physical or functional states. Here I defend reductive realism against three lines of objection derived from Keith Frankish, and I offer two arguments against illusionism. According to one argument, illusionism collapses into realism, and according to the other, it introduces a deep puzzle akin to the hard problem. I conclude that reductive realism is more compelling.*

The idea that phenomenal qualities are merely an illusion strikes some people as patently absurd. I disagree. In fact, there are clinical conditions such as Anton's syndrome that evidently involve self-attribution of experiential states that are not actually occurring. I also think that we are often mistaken about phenomenal qualities. For example, we may be mistaken about which colour experiences, if any, are primary, and there is empirical evidence that we sometimes overestimate the specificity of phenomenal states (for example, we may mistake an array of letter-like forms as actual letters). Moreover, illusionism is less absurd than many more popular competitors, including various forms of dualism that continue to attract adherents. Dualism is patently incompatible with contemporary science, and untestable. It has also spawned a variety of panpsychist views that posit consciousness throughout the cosmos, even though all empirical evidence suggests that it arises under highly circumscribed and predictable circumstances. It is far more plausible that consciousness is an illusion than that it is a pervasive, but non-measurable, feature of the universe. The problem with illusionism is not absurdity. The real problem is that it is outperformed by some reductive realist theories, which identify

Correspondence:
Email: jesse@subcortex.com

phenomenal qualities with functional or physical states. Or so I will argue.

Keith Frankish (this issue) has done a great service by articulating the illusionist agenda. He has clarified the position, and identified key arguments both in its defence and against competing theories. Here I will take on some of those arguments, and also offer some considerations that count against illusionism. I will not attempt to refute Frankish's multi-pronged defence prong by prong. Rather, I'll present a synoptic version of his case against reductive realism, and briefly describe a position that may overcome the challenges that he sets out. Then I'll offer some more general concerns for illusionism, including the charge that it is an unstable position that collapses into realism.

First, a note on terminology. In Frankish's terminology, the position I find most attractive would be called 'weak conservative realism'. 'Conservative' means it is does not require a radical revision or violation of contemporary science, and 'weak' because it allows that we might be partially mistaken about the nature of phenomenal qualities. With all due respect, I'd rather not call my favoured class of theories weak or conservative. This is a reminder that you should never let your opponents name the views you want to defend. First, I prefer the term 'reductive realism' to 'conservative realism', because it draws attention to the fact that such materialist theories do not merely aim for compatibility with science, but work to identify the functional or physical processes that constitute phenomenal experience (plus, I don't really like being called 'conservative'). Frankish's use of the term 'weak' (also something of a pejorative) is meant to emphasize the idea that it is wishy-washy to suppose that we are partially mistaken about phenomenal qualities. He says we can use 'weak realism' and 'weak illusionism' interchangeably, indicating that this position doesn't fall neatly onto either side of the divide. The view that I will defend is not so ambivalent. As noted, I think we may get certain things about phenomenal properties wrong (the status of primary colours, for example), but some of the most controversial properties (such as direct introspectability, intrinsic character, and ineffability) are, I think, defensible. So I will drop the label 'weak', and simply note that reductive realism should make some room for error.

Finally, Frankish describes his illusionism as a thesis about phenomenal consciousness. I prefer to describe it as a thesis about phenomenal qualities, and will also use the term 'qualia' as a synonym. The term 'phenomenal consciousness' implies that there is another kind of consciousness, perhaps what Ned Block calls 'access consciousness'. I side with those who reject this distinction. I think phenomenal properties do not come into existence without being accessible, and I think mere information access is not sufficient for consciousness. Consciousness properly so-called is both phenomenal and accessible. I mention this because it underscores what makes illusionism so controversial. Illusionism is, I believe, a kind of eliminativism about consciousness itself. What we have instead are beliefs about certain sensory states, such as the belief that those states have intrinsic qualities or 'feel like something', and these beliefs, says the illusionist, are false. It would be fairly easy to program a simple computer program to self-attribute such beliefs about its own states. As I understand the position, the illusionist says we are no more conscious than such a program. This is not an absurd thesis, but it is a stunner, if true. The stakes in this debate are high.

I will now present a series of arguments in defence of reductive realism and against illusionism. One view says that certain functional/ physical properties just are phenomenal states and the other says phenomenal states don't exist. How do we decide between these options?

1. In defence of reductive realism

1.1. Can qualia be explained?

David Chalmers has been an enormously helpful contributor to consciousness studies in part because he has a gift for taxonomy. But, in this capacity, he has also done a subtle disservice. He says there are two types of materialism, unhelpfully labelled Type A and Type B (Chalmers, 1997). Type-A materialists say that the science of consciousness will close the explanatory gap that presently exists between scientific theory and the nature of consciousness — that is, qualia can be explained or (as illusionists claim) explained away. Type-B materialists say that the gap will remain after science is complete. Though technically exhaustive, this distinction overlooks a division between two kinds of Type-B materialists: those who think science has nothing to offer by way of explanation (e.g. mysterians), and those who think science can actually explain quite a lot about consciousness. This distinction is important, because the latter view, which I will call partial explanationism, has greater resources for defending realism — resources that bear on Frankish's argument.

In some respects, Frankish reasons more like a dualist than a materialist. He sometimes implies that reductive realism requires explanation: we should be reductive realists only if functional/physical processes can explain phenomenal qualities. He thinks they cannot, and thus we are forced to choose between dualism and illusionism. There is, however, a middle ground. Suppose that science can explain many of the features that are observable in conscious experience but not all. For example, science may be able to explain why after-images are 'negative', why ganzfeld perception leads to a loss of hue, why motion illusions arise and their time course, and so on. Science may even able to explain the minimal threshold of consciousness. We cannot perceive stimuli presented for under 25 milliseconds, which is the duration of a single beat in a gamma wave; if gamma activity is necessary for consciousness, this seemingly arbitrary threshold may have an identifiable source (Prinz, 2012). For partial explanationists, such facts add to the case for reductive materialism. Conscious states are not just correlated with physical states, as dualists concede; they are also to some degree explained by those correlates. Dualism is an ontologically extravagant position that posits entities that go beyond current physics and that cannot be measured using any scientific instrument. These entities are also causally superfluous and mysterious: we can account for behaviour without them, and there is no explanation of how they impact behaviour. Many materialists regard dualism as a non-starter for these reasons. The real question, for such a materialist, is whether the rejection of dualism also entails a rejection of consciousness. If neural states gave us no explanatory purchase on consciousness, this would be a tempting possibility. But, neural states actually account for many properties of consciousness, and this suggests that conscious states do in fact exist and can be located within the nervous system.

Frankish will reply that these neural correlates are just quasi-phenomenal: they are the states we mistakenly describe as phenomenal. But given that these states have many of the properties that we observe in consciousness, why not assume that they also have phenomenal qualities? In other words, partial explanationism (unlike non-explanatory materialism) offers an inductive reason for the identity. In response, Frankish must rely on the dualist's favourite move: if there is something we can't explain physically it must not be part of physical reality. But this principle supposes that science works by deduction. That is indefensible. We cannot derive chemistry from physics or biology from chemistry. Identities are defended in science

1.2. Realism and acquaintance

Frankish knows that many reductionists buy that physical or functional states will not fully explain conscious experience. He discusses the reductionists' favourite strategy for accommodating the explanatory gap: phenomenal concepts. I do not favour this strategy. I don't think we have phenomenal concepts for the simple reason that phenomenal qualities greatly outstrip the range of features we can store in memory, and I think memory, or re-identification, is a criterion for concept possession. I side with those who think we know phenomenal qualities by acquaintance. Gaps arise because acquaintance is not representationally mediated. Phenomenal knowledge is not a description of experience; it *is* experience. Reading neurobiological descriptions of conscious states cannot put us into those states, so it is never adequate for phenomenal knowledge.

Frankish does not discuss acquaintance in the context of the explanatory gap, but he mentions it in another context and raises three objections. These objections can be addressed, if we adopt a materialist theory of acquaintance. First, Frankish says that acquaintance would render phenomenal knowledge psychologically inert, since we need mental representations in order to think about phenomenal qualities. But materialists can rebuff this objection. The states with which we are acquainted are brain states, and, as such, they can influence subsequent information processing. For example, when you taste a wine, you become acquainted with the neural representation of its flavour and hedonic effects. It is these, and not just subsequent beliefs about the wine, that inform your tasting notes, and determine the gusto with which you indulge in your next sip.

The second argument that Frankish gives against acquaintance is that it cannot accommodate the fact that conscious states can be affected by the things we associate with them. I don't see how this follows. The experiences with which we are acquainted are brain states and as such they can be affected by other brain states, including those with which they are associated, and, through back projections, those they effect. Thus, if we associate Sangiovese with strong tannins, for example, we may be more likely to notice tannins in our experience: the association effects the sensory state.

Third, Frankish thinks that acquaintance is a metaphysically dubious construct: 'It is hard to see how physical properties could directly reveal themselves to us' (p. 31). The term 'direct revelation' may sound dubious, but it can actually be applied intelligibly to processes in the nervous system. Three conditions must be met for a state to directly reveal itself: it must qualify as knowledge, it must be known without representational intermediaries, and the knowledge it conveys must include knowledge of the state itself. The neural correlates of conscious states may meet all of these conditions. Consciousness, I noted, always involves accessibility to systems of higher cognition, and, thus, can be regarded as objects of knowledge. On pain of regress, they need not be represented to be known. Unlike the external world, which can be grasped only by mental representations, mental representations (and their neural correlates) constitute our graspings. Finally, unlike many mental states, conscious states inform us about themselves. For example, the neural states that represent colours correspond not to external properties (there are no categorical colour boundaries in the world), but to divisions imposed by the mind. Likewise for the phenomenal representations of so-called primary qualities, such as length. In viewing the Müller-Lyer illusion, we experience one line as longer than the other, but we can't say which line is accurate, since both convey a subjective impression of length, rather than exactly measurable external magnitudes. There is nothing metaphysically mysterious about this. The neural states that constitute experience directly convey how things seem — that is, how they are neurally represented — not necessarily how they actually are. Of course, they don't reveal everything about themselves; they don't reveal that they are neural states, or that they are made of atoms. They reveal only those features that are directly accessible for information processing.

In summary, materialist realists can give a theory of acquaintance, insulating themselves from the dualist arguments that Frankish finds tempting.

1.3. Putatively problematic properties

In defending acquaintance, I suggested that phenomenal states are known directly. In an influential critique of qualia, Dennett (1988) includes direct knowledge, or immediate apprehension, in a list of properties that he takes to be problematic. Others include ineffability, intrinsicness, privacy, and infallibility. Dennett takes these features to be definitive of qualia, and he denies that they are possessed by any mental state; if so, qualia do not exist. Frankish is sympathetic to this argument, and this bolsters his preference for illusionism over reductive realism. I think this argument can be turned on its head: these putatively problematic properties are possessed by certain brain states, at least in certain ways, and this is a reason to prefer reduction over elimination. I have discussed Dennett's argument in more detail elsewhere, so I will be brief here (Prinz, forthcoming).

Phenomenal qualities, I suggested, are known directly, in so far as experience is a kind of knowledge. They are also, for this reason, in some sense infallible. Having phenomenal states is knowing them, so there is no separation between the knowing and the thing known, and thus no room for error. We can, of course, make mistakes when trying to describe our experiences.

What about intrinsicness? This depends on details of the reductive theory. On the theory I prefer, phenomenal states are 'gamma vector-waves': patterns of neural activity oscillating at a particular frequency range (Prinz, 2012). The character of a phenomenal state depends on the wave form, which is an intrinsic property of the state.

Ineffability follows from this picture since the states in question are not verbal in nature. They are not even conceptual. Any effort to describe them would require use of representations at a coarser grain, and thus information would be lost. Ineffability results in a kind of privacy: we cannot directly communicate what these states are like. Moreover, given individual differences in the nervous system, there is no guarantee that any two people with have identical qualia.

One might try to object that there are kinds of privacy, intrinsicness, and so on that cannot be captured on a materialist theory. I think that raises the bar too high. Folk theory attributes these properties, but that doesn't mean folk theory gets them exactly right. Arguments for realism versus elimination work by showing that there are real states that approximate folk theories. I think that condition is met in the case of qualia. Illusionists exaggerate the degree of error.

2. The case against illusionism

2.1. The collapse of illusionism

Having defended reductionist realism against some illusionist objections, I now want to consider some reductionist objections to illusionism. Frankish sometimes describes reductionism (in its weak variety) as an unstable position: once some error is admitted, the road is paved for the view that all beliefs about phenomenal qualities are erroneous. I have granted that we make mistakes about phenomenal qualities, but those arise when we try to represent them in non-phenomenal ways, so the road from partial error to global error is blocked. Now I want to suggest that illusionism is the less stable position. Illusionism may collapse into reductive realism.

There are two ways this might happen. One has to do with reference. We have a vocabulary for talking about phenomenal states: feelings, experiences, qualities, and so on. We learn these words by pointing to examples. Frankish admits that self-ascriptions of phenomenal states have correlates (his 'quasi-phenomenal states'). It follows that these correlates are what we refer to using our phenomenal vocabulary. Quasi-phenomenal sates are phenomenal states, because they are what we point to when assigning meanings to phenomenal vocabulary.

Frankish considers an objection of this kind, and he replies that phenomenal words refer by description rather than by pointing. This reply is not convincing. First, reference by pointing is a common practice, so he owes us a reason for thinking that phenomenal vocabulary refers, in part, by description. Second, the descriptions he has in mind include properties such as ineffability and immediate apprehension. Having just argued that these properties exist, I conclude that phenomenal terms successfully refer, whether by pointing or by description.

The second way illusionism might collapse has to do with the nature of illusions. Illusions arise when things aren't as they seem. Thus, illusions require seemings. For the illusionist, it seems like there is a red quality, but there isn't. But what is a seeming? It is plausible that seeming red just is a red quale. If we replace all qualia talk with talk of how things seem, then we won't have eliminated qualia; we will have redescribed them. Put differently, the gap between seemings and reality can arise only when the feature of reality is not experiential. When the bit of reality in question is an experience, the gap closes, since seemings are experiences.

Frankish could reply that seemings are not experiences. He could insist that they are mere beliefs (or equivalently, he could say, we don't seem to have qualia; we just believe we have qualia). This, however, breaks the analogy with visual illusions, which has been important for this debate. With typical illusions, like the Müller-Lyer, beliefs can be accurate while the illusion persists. The term 'illusionism' might be replaced by 'mistakism'.

A shift from illusionism to mistakism would foreclose one of the two ways in which illusionism collapses into reductionism, but it takes on a burden that is hard for the theory to bear. Beliefs, I have noted, are coarser grained than sensory experiences. Therefore, it's not clear that our faith in qualia could come from false beliefs. After all, the things that seem to exist seem to have qualities that outstrip beliefs. This presents Frankish and other illusionists with a dilemma: either the error arises at a predoxastic state, with seemings, in which case it may collapse into realist reductionism, or the error is doxastic, in which case the seemingly fine grain of experience goes unexplained.

2.2. The illusion problem and the hard problem

These remarks about the source of the error bring us to what Frankish calls the 'illusion problem'. One of the important contributions of his paper is that he lays out an agenda for illusionist theory-building. The illusion problem is at the top of that agenda. It is the problem of explaining how we come to believe we have phenomenal states. Frankish gives a partial answer: evolution may have given rise to the illusion because faith in phenomenal states motivates behaviour in productive ways. The answer is incomplete, however, because we also need a story about how representations of phenomenal states can actually give the impression that we are having phenomenal states.

Notice that mere beliefs about something existing do not generally give rise to corresponding impressions of existence. I believe that there are 'place cells' buried deep in ancient, unconscious regions of brain, but I do not think I experience these. So what is it about beliefs in experience that causes an illusion of experience? Can a belief ever do that without somehow causing an experience? In the opening paragraph, I mentioned Anton's syndrome, a condition in which people believe they are having experiences which they cannot be having. People with this syndrome are blind but think they can see. Here, however, it may be that they are having some kind of sensory experience, such as a visual orienting response, and they simply mistake it for an experience of another kind (a visual perception caused by the external world). Can there be cases where we think we are having an experience and yet we are having none at all?

This question is deep and difficult. The standard analogies for introducing illusionism fail because they do not grapple with the illusion problem adequately. For example, some authors compare the illusion of consciousness to the illusion that colours are out there in the world, rather than generated in the mind. But, in this case, it is normally supposed that there are real experiences, and they are simply getting mislocated. We need a case where there is no experience but there seems to be, and we need an account of how this could arise.

The difficulty of this challenge can be brought out by analogy to the hard problem of consciousness. The hard problem is based on the premise that functional and physical descriptions of the mind-brain do not seem to explain the nature of qualia. The illusion problem may turn out to be hard as well. On the face of it, it seems like magic to suppose that believing an experience exists could give a vivid impression as of an experience. How so? Suppose I volunteer for a masked priming study, and I am told that I unconsciously perceived a red circle followed by a blue circle. I firmly believe what I am told, but I don't have the illusion of experience. I might even learn some trick for knowing when I am presented with one of these colours unconsciously as opposed to the other — perhaps I memorize the sequence or presentation. It still doesn't engender any illusions. Why then does a belief that these states are felt change the situation? Does an illusion arise when I form the belief that my unconscious states are ineffable or immune to error? Unlikely. So we need a story of how functional states, such as beliefs, produce illusions.

Now a suspicion arises: it may be no easier to explain phenomenal illusions then to explain phenomenal consciousness. Indeed, similar resources may be required, such as a special class of concepts or a special form of knowing. But once this is revealed, the attraction of illusionism diminishes. Illusionism purports to dispense with the hard problem, but it leaves us with a parallel problem that may be just as difficult. And this problem is an artefact of the theory, rather than a general problem that every theory must address. It solves one puzzle by introducing another.

3. Conclusions

Illusionism is an attractive option for a materialist. It looks like a lean theory that avoids odd posits like qualia and special epistemic relations, and thus blocks dualism on the cheap. Here I've tried to argue that it may have hidden costs, associated with the illusion problem, and it may collapse into reductive realism. The collapse would not be catastrophic, since I think reductive realists can address the objections that both illusionists and dualists have levied against it. For these reasons, I place my bets with reductionism, but illusionism may be the next best thing. Frankish has done much to raise its plausibility.

References

Chalmers, D.J. (1997) Moving forward on the problem of consciousness, *Journal of Consciousness Studies*, **4** (1), pp. 3–46.

Dennett, D.C. (1988) Quining qualia, in Marcel, A.J. & Bisiach, E. (eds.) *Consciousness in Modern Science*, pp. 42–77, Oxford: Oxford University Press.

Frankish, K. (this issue) Illusionism as a theory of consciousness, *Journal of Consciousness Studies*, **23** (11–12).

Prinz, J.J. (2012) *The Conscious Brain*, New York: Oxford University Press.

Prinz, J.J. (forthcoming) Is consciousness a trick or a treat?, in Huebner, B. & de Brigard, F. (eds.) *The Philosophy of Daniel Dennett*, New York: Oxford University Press.

Georges Rey

Taking Consciousness Seriously — as an Illusion

Abstract: *I supplement Frankish's defence of illusionism by pressing a point I've made elsewhere regarding how actual computational proposals in psychology for conscious processes could be run on desktop computers that most people wouldn't regard as conscious. I distinguish the w(eak)-consciousness of such a desktop from the s(trong)-consciousness people think humans but no such machines enjoy, which gives rise to an explanatory gap, invites first scepticism, unwanted analgesia, and is not supported by Cartesian introspections or any other non-tendentious evidence. Rather, along lines suggested by Wittgenstein and Chomsky, it seems to be an illusory projection of our innate, involuntary responses to things that look and act like our conspecifics, one encouraged by a decontexualized philosophical over-reading of ordinary talk, akin to an over-reading of a term like 'the sky'. However, it also seems to be an illusory projection we can't entirely dispel.*

Keith Frankish has performed a valuable service to consciousness studies by bringing together illusionist views about consciousness (whereby consciousness doesn't exist)[1] that have been too often thoughtlessly dismissed as mad, perverse, or unserious. Some recent philosophers (Chalmers, 1996; Strawson, 2006) are sooner prepared to embrace 'panpsychism', ascribing consciousness to *everything* (whatever that would mean), rather than to doubt it applies to anything. For

Correspondence:
Email: georey2@gmail.com

---

[1] I defend what Frankish (this issue, p. 15) calls 'strong' illusionism, whereby consciousness doesn't exist nowhere, nohow (in particular, not, as 'weak' illusionism proposes, as a real, instantiated state or property about which people simply have false beliefs).

my part, I doubt the relevant issues are to be settled soon.² But it's puzzling that a strategy that has elsewhere often been profitably pursued, of, as Frankish nicely puts it, '[preferring] an explicable illusion to an inexplicable reality' (p. 27), isn't taken more seriously in the case of consciousness. In this short commentary I want to call attention to a number of further issues in light of which I hope the view will receive the more nuanced discussion that Frankish has shown it deserves.

1. S(trong)- vs. w(eak)-consciousness

For the sake of the present discussion, I'll assume a computational theory of many perceptual and cognitive states, attitudes, and processes along lines pursued in a wide swathe of contemporary psychology, including processes of attention, introspection, and higher-order thoughts. Such a system might be said to be 'w(eakly)-conscious', fulfilling many of the conditions we ordinarily take to be sufficient for consciousness — until, I submit, we reflect on the fact that all of these computational processes could be run on a desktop computer that most people wouldn't for a moment be inclined to count as being thereby conscious. I'll say of such people — and I count myself among them — that they have a notion of 's(trong)-consciousness', a notion that seems to me to give rise to Levine's (1983) 'explanatory gap': it seems virtually impossible to imagine what further physical condition a w-conscious desktop should have to satisfy to become s-conscious.³

I've elsewhere tried to make this explanatory gap particularly vivid and plausible by considering what we would say were various computational models proposed by psychologists actually implemented on a desktop computer (equipped, if you like, with a video camera and a robotic arm).⁴ I won't rehearse this possibility here — see Rey

---

[2] 'Panpsychism', though, is a frivolous possibility, for which there's not an iota of evidence; mere worries about reductionism don't begin to support it (see Rey, 2006, for discussion).

[3] Frankish's 'strong' illusionism (see fn. 1) denies the existence of s-, but allows for w-consciousness; his 'weak' illusionism allows for both.

[4] This exercise is particularly important in order to resist what I call 'facile' physicalism, or the view to which I think too many philosophers too readily accede, whereby 'Of course, the mind is a machine, but just an *enormously complex* one'. And, to be sure, it is complex. But one needs to be careful not to replace 'the ghost in the machine' by an insufficiently imagined 'complexity' in it. Imagining running actual psychological

(1995/96) for a discussion of various proposals psychologists and philosophers had made up to that time. I doubt that the argument depends upon the specific proposals I considered there. It's enough merely to look at most any proposal of a computational psychology about some mental phenomenon and to notice how easy it would be to run it on a desktop computer. Indeed, psychologists often test their computational models by running them on machines and testing for not only the same successes, but also, for example, the same errors and response times. For recent example, Peláez and Taniguchi (2016) have along these lines updated the Melzack-Wall model of pain. In the present connection, their final paragraph should be a bit disquieting:

> Our work opens a door in which computational models can be useful for treating pain syndromes: for relieving pain, it is possible to plan a strategy involving plasticity blockers or plasticity enhancers, together with stimulation schedules. This plan can be initially tested in a computer environment so that a personalized strategy can be planned depending on the subject's pain condition. (Peláez and Taniguchi, 2016, §5)

One wonders what they're thinking when they run their model on a computer: if their theory were *true*, wouldn't it imply that the computer would be in whatever pain state they were modelling? Perhaps the model has to be incorporated into the rest of a full psychology: but precisely how and why? Surely vast parts of our psychology are irrelevant to an experience of pain. Notice that the issue can't be evaded by claiming that a model is always different from the phenomena it models: the puzzling question is what would one have to add to an implementation of the model on a computer for it to be in genuine pain, and why they don't consider it in advancing what they take to be an account of pain.

What's at issue here is, of course, a kind of 'zombie' thought experiment. Unlike, however, an appeal to merely the 'conceivability' of two physically identical systems, one of them conscious the other not, where one has no idea of which physical properties might be responsible for which mental ones, in the case I am imagining we are dealing with *actual* proposals about a range of mental phenomena — perception, pain, reasoning, higher-order thought, and decision making — and we are simply asking whether these theories should be

---

proposals on desktops seems to me a valuable way to avoid that error and fasten on what *specifically* is essential to consciousness.

regarded as adequate: if they are, then there should be no problem in regarding the correspondingly programmed desktop to be conscious.[5] If there is a problem, then it would appear that something has been left out. Of course, it's likely there are further computational processes that might well need to be added. However, in discussing the issue over several decades in Europe and America, I've found that this is not the worry audiences usually raise. Overwhelmingly, people feel that something crucial has been left out that is decidedly *not* capturable computationally, but which they also feel at a loss to otherwise specify.[6] I'm, of course, not about to argue that something of this sort *has* been left out. I suspect nothing has been. I'm just struck by the robustness of the intuition that something has, and wonder how this reflects on our — certainly my own — ordinary concept(ion)s of persons and minds.

An analogous situation seems also to arise with regard to free will and personal identity. These notions are also notoriously difficult to capture physically or computationally, at least in ways that can underwrite our usual moral and personal concerns.[7] Here, too, there are 'weak', 'compatibilist' notions that some philosophers have tried to spell out; but many think these inevitably fail, and insist that our ordinary notions are much stronger; and some of these 'incompatibilitists' are quite prepared to find that they may well not in fact apply to actual human beings, a fact that can be made particularly vivid by also considering what it would take to furnish a desktop with the right properties. To Kant's famous trilogy of God, free will, and the soul as 'unschematizable' concepts needed for morality, I'm inclined to add s-consciousness (and delete God).

---

[5] As many have noted, one phenomenon that does resist computational modelling is the ability to integrate information across arbitrary domains (see Fodor, 2000). As interesting as this problem is to the issue of integrative *intelligence*, it would be surprising to find it an obstacle to *consciousness*: the cognitively impaired and many higher animals may lack it, yet still be counted conscious (see Rey, 1996, for discussion).

[6] The exceptions can be counted on one hand, and all are philosophers with a theory to defend. They invariably allow that the desktop is conscious, but 'only to a small degree'; see, for example, Lycan (1996, pp. 39–40). I suppose they are simply concerned with w-consciousness. Of course, if someone really doesn't share the reservations about a w-conscious desktop, there would seem to be little that can be said against them, just as there's little but incredulity to be said against some philosopher's claims of panpsychism. Both seem to me a consequence of the lack of non-tendentious evidence for s-consciousness to be discussed in §4.

[7] For problems with free will, see Strawson (1986); with the ordinary notion of personal identity, Parfit (1971), Nagel (1971), and Schechter (ms).

Note that the computational zombie I'm imagining needn't be the sort imagined by Chalmers (1996, pp. 95–6; quoted in Frankish, this issue, p. 22) for whom 'there is nothing it is like' and for whom 'it is all dark inside'. What it's like for such a creature would be captured by the contents of specific kinds of internal representations that are introspectively available to it, which would involve just as much 'light' as those representations might represent.[8] As Frankish rightly points out, the illusionist may say that 'what it's like'

> depends on a complex of array of introspectable sensory states, which trigger a host of cognitive, motivational, and affective reactions. If we knew everything about these states, their effects, and our introspective access to them, then, illusionists say, we could not clearly imagine a creature possessing them without having an inner life like ours. (p. 23)

Thus, it's open to an illusionist to claim, what seems to me independently plausible, that certain internal states that are tokenings of special representations of pain — for example, ones that are triggered by certain stimulations and in turn trigger strong preferences that the stimulations cease — are every bit as awful as any experience of 'pain itself' would be.[9] As Frankish rightly argues (p. 32), *pace* Searle (1997, p. 112) and Kripke (1980, p. 154), there *can* be an appearance/reality distinction with regard to consciousness: the appearance may simply be of something unreal. Indeed, as Frankish notes (p. 33) and Dennett (1991, p. 81) suggests, first-person authority may amount to no more than the authority an author has over his fictions.

Of course, perhaps a computational account of the relevant mental phenomena *is* inadequate. But, aside from this problem of how we'd react to the account being realized on a desktop, we can presume that there'd be no *scientific* reason to suppose it is. Computational hypotheses about perception and a variety of other mental processes

---

[8] That is, the machine need not be in the position of someone colour-blind who knows all the science of colour perception but not *what it's like* to see red (see Jackson, 1986, and Rey, 1997, chapter 11). Moreover, *pace* both Dennett (1991) and Frankish (p. 34), these introspectable representations may all occur within a computationally unified 'Cartesian theatre' (see Rey, 1994, for discussion).

[9] Here, though, I diverge from Frankish, who wants to allow that 'phenomenal properties are causally potent, considered as intentional objects' (p. 27). That a *representation* of a non-existent intentional object, such as King Lear, a rainbow, or a pain, may be causally efficacious doesn't imply that the non-existent intentional objects '*themselves*' are. Claims about 'intentional inexistents' are simply convenient ways of *classifying* representations, not claims about some further, special realities (see Rey, 2012, for discussion).

seem to be the best explanations of those phenomena on offer. The only alternative to them would seem to have to appeal to some purely physical property of the way in which the computations happen to be implemented in human nervous systems, some further, non-computational property, say, of specific kinds of neurons or neuro-regulators. But it's hard to imagine how such a property might be non-arbitrarily determined.[10] Treating human implementations as somehow privileged in this regard would seem to amount to an extreme form of speciesism, akin to treating certain people as similarly privileged merely because of, say, their skin colour, or whether they have real and not artificial limbs.[11]

2. First-person scepticism and unwanted analgesia

Actually, the situation would be in some ways worse than that, since skin colour and artificiality of limbs might at least be easily ascertained. The further condition that would seem to be needed for s-consciousness would seem to be far more arcane, presently unknown to any of us, and consequently one about which we each might worry whether we have what it takes. Perhaps we only sometimes do; or once did, but do so no longer. How could one know except on arcane authority? Along these lines, Dennett (1975, pp. 219–20) imagines someone who *believes* he is in pain, finds himself averse to torture, and gladly takes aspirin for what he *takes* to be the relief it provides. As Dennett aptly remarks:

> I would not want to take on the task of telling him how fortunate he was to be lacking the *je ne sais quoi* that constituted real pain. (*ibid.*, p. 220)

Indeed, a second consequence of a non-computational condition would be the possibility of a diabolical analgesic: it would be

---

[10] Defenders of 'embodied' and 'extended minds', of course, might suggest that one's body or environment are essential to s-consciousness (see, for example, Clark, 2008). From deference to those with such views, I've imagined the desktop supplied with a video camera and a robotic arm. But what's at issue here seems to me to conflate causal and constitutive dependence: being stabbed in an arm by a knife was doubtless causally crucial for an episode of (a special representation of) pain; but it is nomologically possible for the episode to occur without it.

[11] *Cp.* Block (1978/80) on the tension between overly 'liberal' and overly 'chauvinistic' theories of mind. The famous 'nation of China' example that he discusses there, of course, is another way of bringing up the problem of s-consciousness.

nomologically possible for a patient needing to undergo an extremely painful operation to be deprived of the further, arcane condition while leaving his computational structure entirely intact. There he'd be, under the knife, *believing* himself to be in utterly excruciating pain and as fervently *wishing* not to be as he ever was before being so deprived; but he would be reassured by his, so to say, arcane anaesthesiologist that he was mistaken; he was no more in pain than was the corresponding desktop, which, we may suppose, he had been entirely willing not to regard as consciously suffering any pain. This is, of course, just a way of driving home the point that, on reflection, the appropriate representations of pain can be every bit as awful as any 'real pain' itself.

3. Misplaced Cartesian confidence

The possibility of first-person scepticism should raise an issue that might have been raised *a priori*: what entitles us to any *confidence* we might have in our belief in our own s-consciousness? Descartes famously announced that 'I know plainly that I can achieve an easier and more evident perception of my own mind than of anything else' (1641/1984, I, pp. 22–3), but how does he know the mind he knows is not merely a w-conscious one? It's one thing to be certain of one's w-consciousness, quite another to be certain of one's s-consciousness, and that consequently a suitably programmed machine might not be as conscious as oneself.[12]

It's surely become a commonplace — indeed, it's one of the few achievements of philosophy — that many ideas that seem to most people obvious and undeniable can be shown to be problematic and often false. Naïve views of causality, morality, knowledge, free will, and personal identity have been subject to prolonged and intricate controversies, with scepticism about particularly the latter notions widely taken very seriously. Why should consciousness be any different?

---

[12] How might this problem be raised *a priori*? Once one has divorced propositional attitudes from s-consciousness as Descartes surely should have realized an *explanatory* psychology could do (and it had done in, for example, Platonic views prior to Descartes', and it certainly has recently done), then it follows one could *think* one has s-consciousness without actually having it — one might only have w-consciousness — and so one is not entitled to claim one knows about it incorrigibly.

Descartes, of course, begat a long tradition of attempting to base knowledge on foundations he thought available to consciousness, so that it came to seem to many that knowledge of *anything* depended upon them. But, in the first place, even though Descartes' dualism would suggest that he himself likely had s-consciousness in mind, it's by no means clear that epistemic foundationalism requires it; w-consciousness could equally well suffice. But, secondly, foundationalist approaches in general have been subjected to a variety of criticisms, and can at least no longer be taken for granted, particularly, for many of us, in light of Quine's (1960) 'Neurathian' alternative, whereby any belief can be challenged from the perspective of other beliefs, the lot of them acquiring what warrant they deserve from general considerations of coherence, reliability, and informativeness on the whole.

But perhaps, then, first-person confidence in one's own s-consciousness can be defended on such reliabilist grounds. The case, however, for first-person reliability is not quite as good as Descartes supposed. In their seminal review article regarding introspective reliability, Nisbett and Wilson raised an important possibility that ought at least to give defenders of privileged access to any states pause:

> We propose that when people are asked to report how a particular stimulus influenced a particular response, they do so not by consulting a memory of the mediating process, but by applying or generating causal theories about the effects of that type of stimulus on that type of response... [T]hey may resort in the first instance to a pool of culturally supplied explanations for behavior of the sort in question or, failing in that, to search through a network of connotative relations until they find an explanation that may be adduced as psychologically implying that behavior. (1977, pp. 248–9)

They support this hypothesis by a large range of often surprising experimental examples they reviewed in the article,[13] and conclude about even ordinary examples:

---

[13] E.g. the Hess pupilliary dilation effect, whereby males find photos of female faces more attractive where (unbeknownst to the males) the only difference is the size of the women's pupils; or the Latané 'by-stander' effect whereby people deny what can be shown to be true, that their disinclination to intervene in an emergency had to do with the number of by-standers. There has been much controversy about what to conclude from the many different experiments that Nisbett and Wilson review. One fact that should be stressed right away is that they were concerned about introspection of *causal relations* and *processes*, not perceptual and cognitive *states*. However, the general doubt they raise about introspection seems to me to remain. Wilson (2004) and Carruthers

Thus if we ask another person why he enjoyed a particular party and he responds with 'I liked the people at the party,' we may be extremely dubious as to whether he has reached this conclusion as a result of anything that might be called introspection. We are justified in suspecting that he has instead asked himself Why People Enjoy Parties and has come up with the altogether plausible hypothesis that in general people will like parties because they like the people at the parties. (Nisbett and Wilson, 1977, p. 249)

Nisbett and Wilson stress that none of this implies that the person is wrong, or that there aren't sometimes genuine introspections; only that the process by which a person arrives at what they *take* to be an introspection may sometimes be as informed by popular psychological theories as by any special information that could be said to be genuinely introspected. This is precisely the sort of possibility the illusionist about s-consciousness is raising, which in view of Nisbett and Wilson's and related research can therefore not be lightly dismissed.[14]

4. Lack of non-tendentious data

The only way for the reliability for s-consciousness to be established is for there to be independent evidence of s-consciousness, and for people's introspections of it to be shown to be reliably correlated with it. But this would seem to require evidence for the very reality of s-consciousness that we are seeking.

Frankish (p. 12) makes a useful comparison here with the case of psychokinesis, a purported phenomenon that seems also inexplicable within present physics. A further feature of the comparison that he doesn't mention is the importance of non-tendentious, non-question-begging data in settling on issues of its (un)reality. It's, of course, not

---

(2011) cite further experiments since Nisbett and Wilson's that raise problems about propositional attitudes (although see Rey, 2013, for some reasons to suppose that, *pace* Carruthers, there is *some* genuine introspection of at least *some* of them).

[14] Some philosophers might think that the kind of self-knowledge at issue with s-consciousness is not empirical in the ways that concern Nisbett and Wilson, but more 'reflective' along the lines of knowledge of mathematics. But here they would do well to bear in mind the disturbing case of the brilliant mathematician, John Nash, who was asked (while he was in his schizophrenic phase in a mental hospital), 'How could you, a mathematician devoted to reason and logical proof... how could you believe that extraterrestrials are sending you messages?' He replied: 'Because the ideas about supernatural beings came to me the same way that my mathematical ideas did. So I took them seriously' (Nasar, 1998, p. 11, based on an interview with George Mackey, Harvard, 14 December 1995).

difficult to think of sufficiently controlled experiments that would either show that psychokinesis obtains, or, more likely, that it doesn't. Could similar controlled experiments be imagined for s-consciousness?

This issue is often thought to be a problem for all mental ascription. However, this is certainly not true for most intentional states. Consider the 'standardized tests', of the sort that have been administered in the United States for at least the last eighty years or so, which consist of identically printed sheets of multiple choice 'questions' with corresponding 'answers' provided on a sheet of small rectangles to be filled in with a #2 graphite pencil so that the result can be 'graded' by a machine. Putting aside the 'difficult' questions intended to separate the more from the less talented, the results are statistically staggering. There are millions of striking, counterfactual supporting correlations both among the answer sheets of the respondents and between the answers and the questions asked (see Rey, 1997, chapter 3, for discussion). These corelations can certainly be sufficiently expressed in a non-mentalistic vocabulary — the vast majority of people who were handed the test provided the same graphite patterns — to be both significant and non-tendentious; however, the explanation of them surely requires appeal to intentional states, e.g. the test takers' desires to do well, their understanding of the questions, their abilities to reason, and their intentions to provide what they take to be the correct answers.[15]

Can similar evidence be provided for s-conscious states? Presumably it can be for some w-conscious ones: people do seem to be able reliably to attend and introspect about a good range of their mental states, although, as we've just noted, it's a live empirical question just what the extent of that range may be. But we are imagining that a system could be weakly without being strongly conscious. What further evidence could there be? Perhaps there is some, but it's hard to think of any, especially in light of the fact that reports of it can so often be explained by w-conscious states and illusionist hypotheses alone. In any case, as with any hypothesis, the burden is on the defender of a stronger one to provide the crucial

---

[15] Other non-tendentious data are available. Even more obviously, but with less easily specifiable statistics, are simply cases of people laughing at jokes or weeping at bad news, neither of which could begin to be explained without supposing them to have something like beliefs, desires, and expectations.

evidence.[16] As Frankish notes (p. 26), however, many defenders of s-consciousness have allowed that it might be explanatorily superfluous. By many of their own lights, the best they would be able to do is simply to beat their breasts and declare that they just know 'directly' and 'immediately' that they are s-conscious, and perhaps posit a relation of 'direct acquaintance' with it, ever so tendentiously.[17]

A number of philosophers have claimed that a demand for non-tendentious evidence is unreasonable, since sometimes unsatisfiable, as in the case of, say, the claim that memories are reliable or that anything exists at all, since it would appear that the only evidence we could ever adduce for these claims would be in terms of memories and existing phenomena. The point is perhaps well taken for unusual propositions such as those, although one might think that the fact that their denial would, *à la* Quine's Neurathian conception, wreak havoc with virtually all of our other worldly beliefs provides non-tendentious reason enough in their favour. But, in any case, belief in s-consciousness surely does not remotely play any such crucial role in our thought.[18] Indeed, most claims about s-consciousness could easily be replaced by appeals merely to w-consciousness without damage to much of the rest of our theories of the world. S-consciousness marks a peculiar way in which we think of ourselves and others, a way that it

---

[16] This problem with s-consciousness is one that, of course, Dennett (1991) has also noted. I differ with him over whether one should be equally sceptical about realism with respect to propositional attitudes, for which, as I've argued above, non-tendentious evidence can be provided.

[17] And, of course, the kind of 'acquaintance' relation they would seem to need would be open to the further psychological and metaphysical difficulties Frankish notes (pp. 30–1).

[18] Ned Block (personal correspondence) raised the example of 'Something exists'. A less compelling example is provided by Galen Strawson in an argument explicitly directed at illusionists: 'The experiential realist's correct reply is "It's question-begging of you to say that there must be an account of it that's non-question begging in your terms." Such an exchange shows that we have reached the end of argument, a point further illustrated by the fact that reductive idealists can make exactly the same "You have no non-question-begging account" objection to reductive physicalists that reductive physicalists make to realists about experience' (Strawson, 2006, p. 5, fn. 6).

I'm not sure what 'reductive idealism' Strawson has in mind, but it's simply false that there's no non-tendentious evidence for preferring physics to psychology as a general framework for understanding the world. Is there a psychological explanation that even purports to specify or explain, for example, the nuclear fusion that fuels the stars, in remotely the way that physical/computational theories begin to explain at least w-conscious perceptual phenomena, such as pain and optical illusions? (I might mention that I have found it ironical that Strawson resists illusionism about consciousness, given his very forceful defence of it with regard to free will (1986).)

would be *scientifically coherent* to abandon, even if it would not be either morally or psychologically easy.

5. Other minds

In this latter regard, I think it's important to notice that the problem here is not confined to first-person introspection. Speaking for myself, although I can be as moved as anyone by the standard first-person problems of the apparent immediacy and 'privacy' of s-consciousness, and the seeming undeniability of many of one's own present-tense s-conscious experiences, I am actually just as puzzled by the s-consciousness I regularly attribute to others. At any rate, I seriously find it pretty much as difficult to doubt the s-consciousness of other people and many higher animals as I do my own.

I don't find myself generally sympathetic to Wittgenstein's later views, but I do think he was astute on this point, writing:

> But can't I imagine that people around me are automata, lack consciousness, even though they behave in the same way as usual? If I imagine it now, alone in my room, I see people with fixed looks (as in a trance) going about their business, the idea is perhaps a little uncanny. But just try to hang on to this idea in the midst of your ordinary intercourse with others in the street, say! Say to yourself, for example: 'The children over there are mere automata; all their liveliness is mere automatism.' And you will either find these words becoming quite empty; or you will produce in yourself some kind of uncanny feeling, or something of the sort. (Wittgenstein, 1953/2009, §420)

Indeed, in observing children either playing happily, or crying out in pain, it seems to me easier to imagine that the whole scene is a dream of real, feeling people than that it's real and that they are 'mere automata'. For me, it is pretty much as difficult to imagine of another person being a 'mere machine' of the sort I have sketched as it is of myself, even though I have not the slightest trouble imagining many *aspects* of their and my mental processing being explained by the various sorts of computational theories of perception, reasoning, and decision making psychologists have provided.[19]

---

[19] Of course, someone might explain away my reaction as due to the various complexities of ordinary intelligence normal humans and animals display, many of which have not yet been captured computationally (see fn. 2); and, of course, I can't rule that explanation out *a priori*. But I invite readers to continue the exercise of imagining actual proposals for further capacities being run on a desktop to see if it would really make any difference, and whether they wouldn't feel there to be still an explanatory gap in that

6. A Wittgensteinian diagnosis

Frankish (pp. 5–9) usefully brings together diagnoses illusionists have provided about the sources of the illusion: comparisons with the illusions of impossible figures, apparent images on a computer screen, the apparent world of cartoons, and my own (1995/96) suggestion that it arises from a projection of our innate, involuntary responses in ourselves and to things that look and act like our conspecifics, akin to our projection of secondary properties onto objects that elicit stable responses in us; things that don't elicit these responses, such as a desktop computer, are taken to lack the projected property.

I based this suggestion partly on facts about even young children's automatic responses to, for example, animate movement, upright posture, human faces and speech, and partly on an interesting observation of Wittgenstein:

> [O]nly of a living human being and what resembles (behaves like) a living human being can one say: it has sensations; it sees; is blind; hears; is deaf; is conscious or unconscious. (1953/2009, §281, see also §§282–7, 359–61)

One advanced independently by a diverse number of other writers.[20] I want here to expand on this explanation, conjoining it with another of Wittgenstein's well-known observations about that very gap itself:

> The feeling of an unbridgeable gulf between consciousness and brain process: how come that this plays no role in reflections of ordinary life? ... When does this feeling occur in the present case? It is when I, for example, turn my attention in a particular way on to my own consciousness and, astonished, say to myself: 'this is supposed to be produced by a process in the brain!' as it were clutching my forehead. But what can it mean to speak of 'turning my attention on to my own consciousness'? There is surely nothing more extraordinary than that there should be any

---

case as in the case of the simpler machine. If they don't, fine (see fn. 4). I speak to those like me who do.

[20] See Ziff (1959), Matthews (1977), and Chomsky (2000, p. 44). Note that the condition of being 'living' seems to be what is decisive: so long as computational models are run on things that are 'alive', our reluctance to regard them as s-conscious significantly subsides. Whether this reaction is compatible with a mechanical, non-vitalist conception of 'life' is an issue worth exploring further. I very much suspect it isn't.

Note that I by no means want to endorse the 'living' condition without qualification. As an observation about people's ordinary deployment of the cited mentalistic terms, it seems to be roughly correct; but it also seems to me to wrongly rule out overriding that deployment in the light of scientific theory, which we may need to do to sort out the problem of s- vs. w-consciousness.

such thing! What I described with these words (which are not used in this way in ordinary life) was an act of gazing. I gazed fixedly in front of me, but not at any particular point or object. My eyes were wide open, brows not contracted (as they mostly are when I am interested in a particular object). No such interest preceded this gazing. My glance was vacant; or again, like that of someone admiring the illumination of the sky and drinking in the light. (1953/2009, §412)[21]

I think the way to see the insightful point of this passage is to compare it with our ordinary talk of the 'sky', as when we might say on a sunny, cloudless day that 'The sky is blue'. It's worth spending a few minutes seeing if you can figure out exactly what '*thing*' we even intend to refer to with these terms: just what is the sky and where is it? Is the blue sky here 'the same sky' as that a few miles west where it's raining? Note that when 'the sky is grey' it's of course the clouds '*in it*' that are grey or dark. And is it 'the same sky' in which the stars reside, and into which we rise in an aeroplane, but which isn't the least bit blue 'up close'?[22] What we seem to be doing in saying the various things we do about 'the sky' is engaging in what I think Wittgenstein had in mind as a 'language game'. We all know when and on what basis to use the idiom, but would be taken aback if someone asked us to precisify it and say exactly what the sky was and how and where *it* was blue.

A point with which Wittgenstein would perhaps not be entirely sympathetic is that there is one special language game, that of science and much philosophy, which attempts to explain and characterize the world, as far as possible, independently of human cognition and the varying interests of other language games. Thus, from a scientific point of view, it might be said that, though there is an atmosphere refracting electromagnetic radiation in certain patterns that strike our

---

[21] Note that this is an amended translation from previous editions, including a crucial parenthetical phrase, '(which are not used in this way in ordinary life)', that was strangely omitted in them.

[22] Other examples to think about (and about heroic philosophical efforts to 'define' them) are rainbows, ocean waves, 'voices' (see Dennett, 1968); words, plays, and symphonies; performances, ceremonies, (annulled) marriages, (forged) contracts; the stock market, stock market crashes, companies, stores, clubs; and (to take some now standard examples from Chomsky, 2000, p. 135) flaws in arguments, 'Joe Six Pack', a person's health, and 'the inner track that Raytheon has on the latest missile contract'. I'm indebted to Paul Pietroski for discussion of some of these. See his (2005; 2010) for illuminating development of the underlying Chomskyan conception of linguistic semantics that helps explain these cases, and of the difficulties in identifying actual worldly 'events' as the referents of verbs.

eyes, there really is no such thing as 'the sky', and that it would be a fool's errand to try to make talk of it any clearer or more precise than it ordinarily needs to be. This is, I think, part and parcel of Wittgenstein's general point that philosophers tend to insist that expressions have a precise and determinate application underlying their ordinary use. 'Scientism' might be regarded as the effort to assimilate the uses of all expressions to the scientific language game. In some cases this may be fruitful; in others, pointless. 'The sky is blue' is a nice example of the latter.

Similarly, one ordinarily knows how to use such expressions as 'He has regained consciousness', or 'I was conscious of a certain irony in her remarks', and we pretty clearly do so on the basis of how a normal living thing looks and acts, just as we evaluate 'the sky is blue' by looking up. But, as Wittgenstein goes on to note in the passage, when asked to fasten on consciousness as a phenomenon, we are at a loss to say at all clearly what one is fastening one's attention upon. Indeed, his comparison with drinking in the light of the sky is peculiarly apt: 'consciousness' as it is ordinarily used seems to refer to as undefined and obscure an entity as 'the sky'. And so, again, from a *scientific* point of view, we might similarly deny the existence of consciousness as we would the sky, even though we might, of course, persist in talking 'about them' perfectly well in our ordinary non-philosophical talk.[23]

The point, however, can be over-generalized, as I fear Wittgenstein does in his apparent rejection of *any* scientific explanations of mental processes.[24] *Some* terms and concepts of ordinary language games may well be assimilable into science, as seems to be the case with important aspects of perception, reasoning, language understanding,

---

[23] This Wittgensteinian point can be supplemented by the aforementioned hypothesis of Nisbett and Wilson (1977). It seems to me seriously possible that, as a result of widespread popular philosophizing about the mind, the kind of error Wittgenstein has pointed to could slip onto popular conceptions of the mind, so much so that they are taken to be introspectable.

[24] He dismisses, for example, compensatory adjustment accounts of orientation constancy in vision (1981, §614), empathic understanding of other people (1980, §220), and even *asking* questions about why animals do not talk (1953/2009, §25), whether fishes think (1981, p. 117), or how a shopkeeper knows what to do with the word 'five' — where he famously declares that 'explanations must come to end somewhere' (1953/2009, §1)! In Rey (2003), I discuss at length where Wittgenstein seems to me to go badly wrong in these ways, but then also at the end where he sometimes goes right. The present discussion can be seen as a continuation of the latter enterprise, which, there and here, was influenced by conversations with the late Rogers Albritton.

decision making, and emotions. The more qualified reading of Wittgenstein's strategy seems to me to confine it to ones *where, from a scientific point of view, imaginable research does seem hopeless and pointless in support of the real existence of a phenomenon*, as seems to be the case with the sky and s-consciousness.

7. Persisting puzzles

Let me conclude by conceding that not everything is such smooth sailing for the illusionist as Frankish seems to suggest and I have so far seemed to agree. For I don't think all of the purported illusions have in fact been satisfactorily explained. As Frankish notes (p. 35), Levine (2001, pp. 146–7) called attention to the 'richness and determinacy' of phenomenal experience, a richness and determinacy that in Rey (2007) I have provisionally conceded is not obviously capturable by a computational or other physical theory of mind. I am especially troubled by the old chestnut of phenomenal red/green reversal, which seems at least nomologically possible (see Palmer, 1999). Although both Dennett (1991, pp. 389ff.) and I (1992) have tried to address and defang this possibility, and I think we raised pertinent considerations, I confess to remaining in the end unconvinced by our efforts. Colour experiences seem to me particularly troubling in this regard, since they seem so rich, determinate, and *entirely passive*, without a clear, distinctive computational role to play. I suspect there is here, as in the other phenomenal cases, a certain kind of conceptual illusion, but it is not one that I have yet found a way to explain or dispel, at least not to my (and Levine's) satisfaction. I should confess, however, that I'm partial to even *in*explicable illusions over extravagant, inexplicable realities.

Another, less scientifically troubling, but certainly morally perplexing problem is how at the end of the day to account for the nevertheless persisting distinction I, and I think most people, want to insist upon between human beings, many animals, and ('mere'?) computational equivalents of them. For all of my enthusiasm for a computational theory of mind and for arguments that s-consciousness is an illusion, I find it it is an illusion I can't for the life of me dispel: I simply can't bring myself to think that a suitably programmed desktop computer would have experiences, e.g. of pleasure and pain, red and green, love and remorse, no different from those of a computationally equivalent human being, much less be an object of moral concern. The concept seems to me to be caught up in the rich and intricate ways in which we think about and act towards one another. It is in this way that one can take consciousness quite seriously, even as an illusion. Note the tenacity of the illusion would be in good company. For all my philosophical conviction of the incoherence of our ordinary notions of free will and personal identity, I find I can't seriously give them up either.[25] *Pace* Moore's paradox, certain things seem to me to be true, but I nevertheless can't really believe them.

References

Block, N. (1978/80) Troubles with functionalism, in *Readings in the Philosophy of Psychology*, vol. I, pp. 268–305, Cambridge, MA: Harvard University Press.
Carruthers, P. (2011) *The Opacity of Mind*, Oxford: Oxford University Press.
Chalmers, D.J. (1996) *The Conscious Mind: In Search of a Fundamental Theory*, New York: Oxford University Press.
Chomsky, N. (2000) *New Horizons in the Study of Language and Mind*, Cambridge: Cambridge University Press.
Clark, A. (2008) *Supersizing the Mind: Embodiment, Action, and Cognitive Extension*, New York: Oxford University Press.
Dennett, D. (1968) *Content and Consicousness*, London: Routledge.
Dennett, D. (1975) Why you can't make a computer that feels pain, in *Brainstorms*, pp. 190–229, Cambridge, MA: MIT Press.
Dennett, D. (1991) *Consciousness Explained*, New York: Little, Brown.
Descartes, R. (1641/1984) Meditations on first philosophy, in Cottingham, J., Stroothoff, R. & Murdoch, D. (eds.) *The Philosophical Writings of Descartes*, Cambridge: Cambridge University Press.
Fodor, J. (2000) *The Mind Doesn't Work That Way*, Cambridge, MA: MIT Press.
Jackson, F. (1986) What Mary didn't know, *The Journal of Philosophy*, **83** (5), pp. 291–295.
Kripke, S. (1980) *Naming and Necessity*, Cambridge, MA: Harvard University Press.
Levine, J. (1983) Materialism and qualia: The explanatory gap, *Pacific Philosophical Quarterly*, **56**, pp. 354–361.
Levine, J. (2001) *Purple Haze: the Puzzle of Consciousness*, New York: Oxford University Press.
Lycan, W. (1996) *Consciousness and Experience*, Cambridge, MA: MIT Press.
Matthews, G. (1977) Consciousness and life, *Philosophy*, **52** (January), pp. 13–26.
Nagel, T. (1971) Brain bisection and the unity of consciousness, *Synthese*, **22**, pp. 396–413.
Nasar, S. (1998) *A Brilliant Mind: The Life of Brilliant Mathematician and Nobel Laureate, John Nash*, New York: Touchstone.
Nisbett, R. & Wilson, T. (1977) Telling more than we can know: Verbal reports on mental processes, *Psychological Review*, **84** (3), pp. 231–259.

---

[25] Come to think of it, I also have trouble in my heart of hearts giving up the sky, rainbows, and the reality of secondary properties. Simply incorrigible, I am (though not in the way Cartesians have sometimes claimed).

Palmer, S. (1999) Color, consciousness, and the isomorphism constraint, *Behavioral and Brain Sciences*, **22** (6), pp. 1–21.
Parfit, D. (1971) Personal identity, *Philosophical Review*, **80**, pp. 3–27.
Peláez, F. & Taniguchi, S. (2016) The gate theory of pain revisited: Modeling different pain conditions with a parsimonious neurocomputational model, *Neural Plasticity*, [Online], http://dx.doi.org/10.1155/2016/4131395.
Pietroski, P. (2005) Meaning before truth, in Preyer, G. & Peters, G. (eds.) *Contextualism in Philosophy*, pp. 253–300, Oxford: Oxford University Press.
Pietroski, P. (2010) Concepts, meanings, and truth: First nature, second nature, and hard work, *Mind and Language*, **25**, pp. 247–278.
Quine, W. (1960) *Word and Object*, Cambridge, MA: MIT Press.
Rey, G. (1992) Sensational sentences switched, *Philosophical Studies*, **67**, pp. 73–103.
Rey, G. (1994) Dennett's unrealistic psychology, *Philosophical Topics*, **22** (1–2), pp. 259–289.
Rey, G. (1995/96) Annaherung an eine projektivistische Theorie bewuBten Erlebens, in Metzinger, T. (ed.) *BewuBtsein*, Paderborn: Ferdinand-SchöninghVerlag. English version (1996) Towards a projectivist account of conscious experience, in Metzinger, T. (ed.) *Conscious Experience*, pp. 123–142, Paderborn: Ferdinand-Schöningh-Verlag.
Rey, G. (1997) *Contemporary Philosophy of Mind: A Contentiously Classical Approach*, Oxford: Blackwell.
Rey, G. (2003) Why Wittgenstein ought to have been a computationalist (and what a computationalist can learn from Wittgenstein), *Croatian Journal of Philosophy*, **III** (9), pp. 231–264.
Rey, G. (2006) Better to study human than world psychology, *Journal of Consciousness Studies*, **13** (10–11), pp. 110–116.
Rey, G. (2007) Phenomenal content and the richness and determinacy of color experience, *Journal of Consciousness Studies*, **14** (9–10), pp. 112–131.
Rey, G. (2012) Externalism and inexistence in early content, in Schantz, R. (ed.) *Prospects for Meaning*, pp. 503–529, New York: deGruyter.
Rey, G. (2013) We aren't all self-blind: A defense of a modest introspectionism, *Mind and Language*, **28** (3), pp. 259–285.
Schechter, E. (ms) *The Other Side: Self-consciousness and 'Split' Brains*.
Searle, J. (1997) *The Mystery of Consciousness*, New York: A New York Review Book.
Strawson, G. (1986) *Freedom and Belief*, Oxford: Clarendon Press.
Strawson, G. (2006) Realistic monism: Why physicalism entails panpsychism, in Freeman, A. (ed.) *Consciousness and its Place in Nature: Does Physicalism Entail Panpsychism?*, pp. 3–31, Exeter: Imprint Academic.
Wilson, T. (2004) *Strangers to Ourselves*, Cambridge, MA: Harvard University Press.
Wittgenstein, L. (1953/2009) *Philosophical Investigations*, Anscombe, G.E.M., Hacker, P.M.S. & Schulte, J. (trans.), revised 4th ed., Hacker, P.M.S. & Schulte, J., Oxford: Blackwell.
Wittgenstein, L. (1980) *Remarks on the Philosophy of Psychology*, II, Anscombe, G.E.M. & von Wright, G.H. (eds.), Oxford: Blackwell.
Wittgenstein, L. (1981) *Zettel*, Anscombe, G.E.M. & von Wright, G.H. (eds.), Oxford: Blackwell.
Ziff, P. (1959) The feelings of robots, *Analysis*, **19**, pp. 64–68.

Amber Ross

Illusionism and the Epistemological Problems Facing Phenomenal Realism

Abstract: *Illusionism about phenomenal properties has the potential to leave us with all the benefit of taking consciousness seriously and far fewer problems than those accompanying phenomenal realism. The particular problem I explore here is an epistemological puzzle that leaves the phenomenal realist with a dilemma but causes no trouble for the illusionist: how can we account for false beliefs about our own phenomenal properties? If realism is true, facts about our phenomenal properties must hold independent of our beliefs about those properties, so mistaken phenomenal beliefs must always remain an open possibility. But there is no way to identify the phenomenal facts that make these beliefs false other than by mere stipulation. If illusionism is true, then the state of affairs regarding what a subject's experience seems like is just the illusion itself; there are no further facts of the matter about which the subject might have mistaken beliefs, so the problem does not arise.*

Phenomenal realism, whether radical or conservative, appears to be the default position regarding conscious experience. Anything less than full-fledged realism seems to deny the existence of our subjective inner lives. But the proposal that Keith Frankish gives us here, a type of anti-realism he calls 'illusionism' about phenomenal properties, has the potential to leave us with all of the benefit of taking conscious experience seriously and much less of the unnecessary metaphysical and conceptual baggage that accompanies a realist interpretation of these properties.

Correspondence:
Amber Ross, University of Toronto, Canada.   *Email: amber.ross@utoronto.ca*

In categorizing phenomenal properties as illusions, it is important to note that Frankish is not suggesting that our inner lives are any less rich than they seem to be. On the contrary, our conscious experience is exactly as rich as it seems to be, not because there are real phenomenal properties present in our experience but because being rich and full of feeling is precisely how our experience seems to us, how our introspective representational mechanisms present our experiences to us. Frankish takes issue with the characterization of a philosophical zombie — a creature physically identical to an ordinary human being — as something 'with no inner life, whose experience is completely blindsighted' (Frankish, this issue, p. 22). The 'inner life' of any creature whose introspective representational mechanisms present their experience as rich and full of feeling *just is* rich and full of feeling. According to Frankish, 'having the kind of inner life we have… consists of having a form of introspective self-awareness that creates the illusion of a rich phenomenology' (*ibid.*). If we take illusionism seriously, then there is no further question of whether we are accurately representing *real* (rather than ersatz or pseudo) phenomenal properties. Zombies are just as correct in their judgments about how their experience *seems to them* as their ordinary human counterparts.

As difficult as it might be to come to terms with the idea that phenomenal consciousness is a type of illusion, the epistemological problems facing the phenomenal realist may actually outweigh those facing illusionism. As Frankish aptly notes, 'we have no introspective way of checking the accuracy of our introspective representations, and so cannot rule out the possibility that they are non-veridical' (p. 28). He mentions this in the context of an argument for taking illusionism seriously, but there are multiple ways in which our introspective representations could turn out to be non-veridical, and potentially the most complicated arise for realist views. If illusionism is correct, then our introspective representations are non-veridical in so far as we interpret them as representing real properties rather than mere 'intentional objects, or a sort of mental fiction' (*ibid.*). If realism is correct, certain failures of representation must still be a possibility. However, the manner in which our introspective representations can turn out to be non-veridical is more complicated: while the existence of phenomenal properties would depend upon the existence of our conscious experience, if phenomenal properties are real then their nature is necessarily independent of our beliefs about our experience.

Those who admit phenomenal properties into their ontology encounter serious epistemic difficulties, regardless of whether their underlying metaphysics is physicalist or dualist. To use Frankish's terms, both 'radical' and 'conservative' phenomenal realists face challenging puzzles regarding our epistemic relation to the phenomenal properties of our experience. To be a realist about phenomenal properties one must be able to defend the possibility of a scenario in which a subject believes herself to have a conscious experience with a particular phenomenal character while she is actually having a conscious experience with a different phenomenal character. That is, subjects must be able to genuinely hold mistaken beliefs about the content of their own conscious experience, about *what it's like* to be them at that moment. Cases in which subjects are not closely attending to their experience, or in which they misremember previous experiences and make inaccurate comparisons between current conscious experiences and prior ones, are easy to conceive. But if phenomenal realism is true, then subjects must be able to make mistakes about the content of their conscious experiences even when they are attending carefully, and even when their beliefs only concern their current conscious experiences. If a property is real, there are objective facts about that property, which is to say that whatever is true about these properties will be true regardless of a subject's beliefs about them. So a subject must be able to hold mistaken beliefs about the intrinsic (rather than relational) content of their occurrent conscious experience. Any treatment of phenomenal properties that falls short of this will fail to qualify as genuine realism, and will rather be some sort of disguised illusionism.

Constructing a scenario in which a subject believes herself to have a conscious experience with phenomenal character Φ (that is, a conscious experience that instantiates phenomenal property Φ) while she is actually having an experience with phenomenal character Ψ is fairly difficult. Chalmers (2003) attempts to do so when discussing the epistemology of phenomenal belief, and examining the scenario he constructs will illuminate certain epistemological challenges facing the phenomenal realist, challenges an illusionist will be able to avoid. To provide a framework for conceptualizing phenomenal belief — beliefs about our own occurrent conscious experiences — Chalmers coins a set of technical terms. One is the notion of a *direct phenomenal concept*, a concept the subject deploys via introspection which 'by its nature picks out instances of an underlying demonstrated phenomenal quality' (*ibid.*, p. 242) or phenomenal property. When

Mary sees colour for the first time and has the thought, 'Oh, so that's *what it's like to see red*', the concept we could articulate as 'what it's like to see red' which she deploys here is what Chalmers calls a 'direct phenomenal concept'.[1] Direct phenomenal concepts partially constitute *direct phenomenal beliefs*, the kind of belief that 'identifies the referent of that concept with the very demonstrated quality' (*ibid.*), as Mary's belief above identifies the demonstrated phenomenal quality (the referent of 'that') with the referent of the direct phenomenal concepts ('what it's like to see colour'). So a direct phenomenal belief is formed when a subject internally gestures towards an experience she is having at time t_1 with an introspective token-demonstrative '*this (Q)... experience*', and this kind of belief lasts only as long as the subject attends to the experience.[2] If phenomenal properties are real, whether they are ultimately physical or non-physical, it should be possible for subjects to hold direct phenomenal beliefs, since these beliefs, when justified and true, constitute our phenomenal knowledge.[3] And it is this sort of phenomenal belief that subjects must be able to hold in error, since the truthmaker for such a belief is not how a subject's conscious experience seems to her but rather the facts about the phenomenal properties instantiated in her conscious

---

[1] It would be more appropriate to characterize the direct phenomenal concept in Mary's belief as 'what it's like to *have this experience*' rather than 'what it's like to *see red*', since characterizing it as a *red experience* introduces a relational element we are trying to avoid here, in particular a connection between the experience she is having and what she already knows about colour phenomena, in particular, that some stimuli evoke red responses.

[2] Since both direct phenomenal beliefs and direct phenomenal concepts last only as long as a subject attends to a particular experience, there are some views of concepts according to which direct phenomenal concepts will not qualify as concepts, and will leave the status of Mary's belief an open question. For example, per the definition of 'concept' from Prinz (2007) and Millikan (2000), for something to be a concept it must be redeployable; that is, it must be the kind of thing a subject can deploy on multiple occasions, as the content is determined or fixed only by similarities between various instances in which a subject is disposed to deploy the concept. So by merely attending to an experience a subject is having at time t_1 and internally gesturing toward it with a introspective token-demonstrative '*this (Q)... experience*', the subject will not have formed a phenomenal concept. On the Millikan/Prinz view, *Q* here cannot be a concept unless it can be reused to identify multiple instances of the same phenomenal state, and this would conflict with the stipulated definition of direct phenomenal concepts.

[3] While there is room in logical space for a position that embraces realism about phenomenal properties but rejects the possibility of phenomenal knowledge, such a position would lack any intuitive or philosophical appeal.

experience, which are at least conceptually if not metaphysically independent of her beliefs about the character of her experience.

The 'clear case' of failed direct phenomenal belief that Chalmers (2003) attempts to provide is perhaps as close as one can come to presenting such a case, but it is far from obvious that it actually constitutes a genuine case of this type. A direct phenomenal belief is only formed when a subject deploys a demonstrative phenomenal concept and a direct phenomenal concept 'based in the same act of attention' (*ibid.*, p. 236) and lasts only as long as that act of attention. For a subject to successfully form a direct phenomenal belief, her demonstrative phenomenal concept and direct phenomenal concept must be 'appropriately aligned' (*ibid.*). A subject will attempt but fail to form a direct phenomenal belief (or will form a false direct phenomenal belief) in cases where her demonstrative act fails to pick out a referent. In such a case, a subject intends to attend to some phenomenal property instantiated in her phenomenal experience and (in so doing) to form a direct phenomenal concept of that property, but somehow fails in this attempt. Chalmers attributes this failure to a 'mismatch' between the cognitive element of the demonstration and targeted experiential element, i.e. the phenomenal property.

Chalmers' example is the case of mildly-misfortunate Nancy, who attends to a coloured patch of her phenomenal field, acting cognitively as if to demonstrate a highly specific phenomenal shade. Nancy intends to attend to her phenomenal experience in such a way that she would demonstrate a patch of phenomenal colour; that is, the object of Nancy's attention and intended demonstration — the patch of phenomenal colour — is a phenomenal property. But her demonstration fails, purportedly because the cognitive elements and targeted experiential elements of her attempted demonstration are 'mismatched'. In Chalmers' thought experiment,

> ...Nancy has not attended sufficiently closely to notice that the patch has a nonuniform phenomenal color: let us say it is a veridical experience of a square colored with different shades of red on its left and right side... (*ibid.*, p. 237)

If we consider this scenario for a moment, we will see that there are at least two stipulated features of Nancy's situation that make her 'false belief' seem plausible, both of which are suspect.[4] First, the notion

---

[4] We might choose to say that Nancy has failed to form a direct phenomenal belief, rather than attributing to her a full-fledged false belief. For our purposes here either

that Nancy's phenomenal field could be differently shaded on its left and right side while she believes it to be uniformly coloured seems *prima facie* plausible when it is described as a veridical experience of something in the external world, the properties of which can be objectively confirmed. But we should ask what *reason* we have for positing that Nancy's phenomenally conscious experience is a veridical representation of this multi-shaded square when this assertion conflicts with the subject's own belief about how her phenomenal field seems to her. To Nancy, it seems as if her phenomenal field has a consistent colour throughout, and without any way to confirm the 'phenomenal facts' regarding her conscious experience, to claim that the phenomenal content of her experience is anything other than how she takes it to be would be mere stipulation without support. Perhaps connecting the facts about Nancy's phenomenal field to states of affairs in the external world could provide suitable support, but since we are concerned solely with her subjective phenomenal experience, any connection between her phenomenal field and the world it represents is irrelevant, as Chalmers subsequently seems to agree. A later version of this material appears in Chalmers (2010, chapters 8 and 9), which are a near reprint of Chalmers (2003). In his subsequent account of Nancy's failed phenomenal belief, the qualifier 'veridical' is omitted from the description of the content of her phenomenal experience. Instead, the passage reads: 'Let us say that she has an experience of a square colored with different shades of red on its left and right side' (2010, p. 270). And rightly so, as the use of 'veridical' is unlicensed here, an unsupportable stipulation. But if the realist cannot appeal to the notion of 'veridical representation' to specify the content of Nancy's experience, he loses the intuitively plausible grounds for his claim that the *actual* content of the subject's phenomenal experience and her *beliefs* about her phenomenal experience are 'mismatched'. Her failure is left unexplained.

Stipulating that Nancy's beliefs about her experience do not accurately represent how her experience seems to her flies in the face of our intuitive conception of our relation to our own conscious experience, a problem the illusionist can easily avoid. If illusionism is true, a subject may still misjudge a current quasi-phenomenal red

---

interpretation would be adequate: Nancy has made an odd mistake in judgment regarding the phenomenal properties present in her conscious experience, which has resulted in either a false belief or a failed attempt at forming such a belief.

experience as being identical to a past quasi-phenomenal red experience. But if Frankish is correct and there are no real phenomenal properties, then a subject's judgments about how her current experiences seem to her now (rather than how they relate to other experiences) cannot fail to match some fact of the matter about the *real* phenomenal properties of her experience. The realist, however, must be willing to accept that a subject's beliefs may always misrepresent the facts of the matter about their phenomenal experiences. He is committed to the notion that the facts about the phenomenal character of a subject's conscious experience are set by facts about the real phenomenal properties instantiated in her experience, and part of what makes the view *realism* is that those facts are in no way determined by a subject's beliefs about how her experience seems to her. Any view according to which the subject's beliefs about the character of her conscious experience do play a role in determining the facts of the matter about her conscious experience is a non-realist, illusionist type of view.

The second feature of this situation that makes Nancy's false belief seem plausible is that Nancy is described as not attending closely to her experience. The only circumstance Chalmers acknowledges in which a subject can attempt but fail to form a direct phenomenal belief is when the subject is not adequately attending to her experience (see Chalmers, 2003). As Frankish points out (p. 30), a radical or anti-physicalist realist may appeal to 'acquaintance' as our form of epistemic access to the phenomenal properties of our experience, so long as we are adequately attending to those properties (see Chalmers, 1996, pp. 196–7; 2003, pp. 246–54). And, as Frankish notes, the acquaintance relation is no magic bullet, even for anti-physicalists, since we must mentally represent our phenomenal properties in order to think or talk about them, and the representational process is by its nature potentially fallible (p. 30). If phenomenal realism is true, then Nancy must be able to form the false belief that her phenomenal field is consistently coloured when it is actually non-uniform, regardless of how closely she is attending to her experience. Her belief represents the character of her conscious experience, it does not constitute the character of that experience, and representation always carries the possibility of misrepresentation. Even if we grant that there is some sort of certainty guaranteed by the acquaintance relation, this certainty will not transfer to our beliefs about the phenomenal character of our conscious experience. If phenomenal properties are real, we must be capable of making errors in our judgment about their nature regardless of how closely we attend to the content of our experience. As long as the phenomenal and cognitive components of our mental states remain distinct (either conceptually or metaphysically), the cognitive effort we put forth in forming those beliefs will provide no guarantee that our beliefs will accurately represent the facts about our phenomenal properties. But to say that Nancy attends as closely as possible to the phenomenal character of her conscious experience and still forms false beliefs about it seems to leave us at an awkward distance from the content of our own experience, a distance that the phenomenal realist is committed to embracing.

The phenomenal realist is burdened with the problem of finding a plausible way to answer the question of whether a subject's phenomenal beliefs satisfy the criteria for phenomenal knowledge. Not all phenomenal beliefs will succeed here; there must be room for some phenomenal beliefs to turn out false. Hence the realist encounters a serious *problem of one's own mind*. An illusionist has no such problem, and faces no analogous challenge; subjects hold phenomenal beliefs, but there is no further question of whether these beliefs fit the criteria for phenomenal knowledge, because according to the illusionist phenomenal beliefs do not represent real properties — there is nothing against which their 'accuracy' can appropriately be measured. Phenomenal belief is the whole story, and there is nothing to be gained by searching for further 'phenomenal facts' nor by asking whether those beliefs amount to 'phenomenal knowledge'. If the illusionist is correct, the answer to this question is 'no'; there is no such thing as phenomenal knowledge in the sense advocated by the phenomenal realist.

For the phenomenal realist, the standards against which phenomenal beliefs can be deemed true or false are set by phenomenal facts — facts about one's own phenomenal states that are independent of one's beliefs about those states. For the radical (anti-physicalist) realist, these facts will be independent of any physical facts about the subject. For the conservative realist, phenomenal facts will be a special subset of physical facts, but must still be (either conceptually or in actuality) independent of a subject's phenomenal beliefs. As Frankish rightly points out, conservative realism balances on a knife edge, and if it turns out that, though we may have phenomenal beliefs (beliefs about how our experience seems to us), there is no such thing as an actual phenomenal fact, the conservative position will collapse into illusionism.

References

Chalmers, D.J. (1996) *The Conscious Mind: In Search of a Fundamental Theory*, New York: Oxford University Press.

Chalmers, D.J. (2003) The content and epistemology of phenomenal belief, in Smith, Q. & Jokic, A. (eds.) *Consciousness: New Philosophical Perspectives*, pp. 220–272, Oxford: Oxford University Press.

Chalmers, D.J. (2010) *The Character of Consciousness*, New York: Oxford University Press.

Millikan, R. (2000) *On Clear and Confused Ideas*, Cambridge: Cambridge University Press.

Prinz, J. (2007) Mental pointing: Phenomenal knowledge without concepts, *Journal of Consciousness Studies*, **14** (9–10), pp. 184–211.

Eric Schwitzgebel

Phenomenal Consciousness, Defined and Defended as Innocently as I Can Manage

Abstract: *Phenomenal consciousness can be conceptualized innocently enough that its existence should be accepted even by philosophers who wish to avoid dubious epistemic and metaphysical commitments such as dualism, infallibilism, privacy, inexplicability, or intrinsic simplicity. Definition by example allows us this innocence. Positive examples include sensory experiences, imagery experiences, vivid emotions, and dreams. Negative examples include growth hormone release, dispositional knowledge, standing intentions, and sensory reactivity to masked visual displays. Phenomenal consciousness is the most folk-psychologically obvious thing or feature that the positive examples possess and that the negative examples lack, and which preserves our ability to wonder, at least temporarily, about antecedently unclear issues such as consciousness without attention and consciousness in simpler animals. As long as this concept is not empty, or broken, or a hodgepodge, we can be phenomenal realists without committing to dubious philosophical positions.*

1. Introduction

Keith Frankish argues that phenomenal consciousness does not really exist. I, along with most other Anglophone philosophers who have written on the issue, think that phenomenal consciousness does exist.

Frankish can be interpreted as posing a dilemma for defenders of the real existence of phenomenal consciousness — *phenomenal realists*,

Correspondence:
Eric Schwitzgebel, Department of Philosophy, University of California at Riverside, Riverside, CA 92521-0201, USA.
Email: eric.schwitzgebel@ucr.edu

as I will call them. On Horn 1 we find inflated views about what phenomenal consciousness involves: infallibility, or metaphysical dualism, or some other dubious philosophical commitments. If phenomenal consciousness requires any of *those* features, it probably doesn't exist. On Horn 2 we find views so deflationary as to be tantamount to the non-existence of the originally intended phenomenon. Frankish argues that we must choose between thinking of phenomenal consciousness in so inflated a way that there fails to be any such thing or so deflationary a way that it has effectively vanished into something else (e.g. dispositions to make certain sorts of judgments).

The best way to meet Frankish's challenge is to provide something that the field of consciousness studies in any case needs: a clear definition of phenomenal consciousness, a definition that targets a phenomenon that is both *substantively interesting* in the way that phenomenal consciousness is widely thought to be interesting but also *innocent of problematic metaphysical and epistemological assumptions*. In Section 2, I will attempt to do this.

One necessary condition of being substantively interesting in the relevant sense is that phenomenal consciousness should retain at least a superficial air of mystery and epistemic difficulty, rather than collapsing immediately into something as straightforwardly deflationary as dispositions to verbal report, or functional 'access consciousness' in Block's (1995/2007) sense, or an 'easy problem' in Chalmers' (1995) sense. If the reduction of phenomenal consciousness to something physical or functional or 'easy' is possible, it should take some work. It should not be *obviously* so, just on the surface of the definition. We should be able to wonder how consciousness could possibly arise from functional mechanisms and matter in motion. Call this the *wonderfulness* condition.

2. Defining consciousness by example

Unfortunately, the three most obvious, and seemingly respectable, approaches to definition all fail. Phenomenal consciousness cannot be defined *analytically*, in terms of component concepts (as 'rectangle' might be defined as a right-angled planar quadrilateral). It is a foundationally simple concept, not divisible into component concepts. Even if it were to prove upon sufficient reflection to be analytically decomposable without remainder, *defining* it in terms of some hypothesized decomposition, right now, at our current stage of enquiry, would beg the question against researchers who would reject such a decomposition. Widespread disagreement also makes it inadvisable to define phenomenal consciousness *functionally*, in terms of the causal role it normally plays (as 'heart' might be defined as the organ that normally plays the causal role of pumping blood). It's too contentious what causal role, if any, phenomenal consciousness might have. Nor, for present purposes, can phenomenal consciousness be adequately defined by *synonymy*, since Frankish's inflation-or-deflation dilemma applies equally to all of the nearby terms and phrases like 'qualia', 'what-it's-like-ness', or 'stream of experience'.

The best approach is *definition by example*. Definition by example can sometimes work well, if one provides diverse positive and negative examples and if the target concept is natural enough that the target audience can be trusted to latch onto that concept once sufficient positive and negative examples are provided. I might say 'by *furniture* I mean tables, chairs, desks, lamps, ottomans, and that sort of thing; and not pictures, doors, sinks, toys, or vacuum cleaners'. Hopefully, you will latch on to approximately the relevant concept (e.g. not being tempted to think of a ballpoint pen as furniture but being inclined to think that a dresser probably is). I might even define *rectangle* by example, by sketching out for you a variety of instances and nearby counter-instances (triangles, parallelograms, trapezoids, open-sided near-rectangles). Hopefully, you get the idea.

Definition by example is a common approach among recent phenomenal realists. I interpret Searle (1992, p. 83), Block (1995/2007, p. 166–8), and Chalmers (1996, p. 4) as aiming to define phenomenal consciousness by a mix of synonymy and appeal to example, plus maybe some version of the wonderfulness condition. All three attempts are, in my view, reasonably successful. However, all three attempts also have three shortcomings, which I aim to repair here. First, they are not sufficiently clear that they *are* definitions by example, and consequently they don't sufficiently invite the reader to reflect on the conditions necessary for definition by example to succeed. Second, perhaps partly as a result of the first shortcoming, they don't provide enough of the negative examples that are normally part of a good definition by example. Third, they are either vague about the positive examples or include needlessly contentious cases. Siewert (1998, chapter 3) is somewhat clearer on these points, but still limited in his range of negative examples and in his exploration of the conditions of failure of definition by example.

I want to highlight one crucial background condition that is necessary for definition by example to succeed. There must be an *obvious* or *natural* category or concept that the audience will latch onto once sufficiently many positive and negative examples have been provided. In defining *rectangle* by example for my eight-year-old daughter, I might draw all of the examples with a blue pen, placing the positive examples on the left and the negative examples on the right. In principle, she might leap to the idea that 'rectangle' refers to retangularly-shaped-things-on-the-left, or she might be confused about whether red figures can also be rectangles, or she might think I am referring to spots on the envelope rather than to the drawn figures. But that's not how the typical enculturated eight-year-old human mind works. The definition succeeds because I know she'll latch onto the intended concept, rather than some less obvious concept that fits the cases.

Defining *phenomenal consciousness* by example requires that there be *only one* obvious or readily adopted concept or category that fits with the offered examples. I do think that there is probably only one obvious or readily adopted category in the vicinity, at least once we do some explicit narrowing of possible candidates. In Section 3, I will discuss concerns about this assumption.

Let's begin with positive examples. The word 'experience' is sometimes used non-phenomenally (e.g. 'I have twenty years of teaching experience'). However, in normal English it often refers to phenomenal consciousness. Similarly for the adjective 'conscious'. I will use those terms in that way now, hoping that when you read them they will help you latch onto relevant examples of phenomenal consciousness. However, I will not always rely on those terms. They are intended as aids to point you towards the examples rather than as (possibly circular or synonymous) components of the definition.

Sensory and somatic experiences. If you aren't blind and you think about your visual experience, you will probably find that you are having some visual experience right now. Maybe you are visually experiencing black text on a white page. Maybe you are visually experiencing a computer screen. If you press the heels of your palms firmly against your closed eyes for several seconds, you will probably notice a swarm of bright colours and figures, called phosphenes. All of these visual goings-on are examples of phenomenal consciousness. Similarly, you probably have auditory experiences if you aren't deaf — at least when you stop to think about it. Maybe you hear the hum of your computer fan. Maybe you hear someone talking down the hall. In a sufficiently quiet environment, you might even hear the rush of blood in your ears. If you cup one hand silently over one ear, you will probably notice the change in ambient sound. If you stroke your chin with a finger, you will probably have tactile experience. Maybe you are feeling the pain of a headache. If you sip a drink right now, you will probably experience the taste and feel of the drink in your mouth. If you close your eyes and think about where your limbs are positioned, you might have proprioceptive experience of your bodily posture, which you might notice becoming vaguer if you remain motionless for an extended period.

Conscious imagery. Maybe there's unconscious imagery, but if there is, it's doubtful that you will be able to reflect upon an instance of it at will. Try to conjure a visual image — of the Eiffel Tower, say. Try to conjure an auditory image — of the tune of 'Happy Birthday', for example. Imagine it sung in your head. Try to conjure a motor image. Imagine how it would feel to stretch your arms back and wiggle your fingers. You might not succeed in all of these imagery tasks, but hopefully you succeeded in at least one, which you can now think of as another example of phenomenal consciousness.

Emotional experience. Presumably, you have had an experience of sudden fear on the road, during or after a near-accident. Presumably, you have felt joy, surprise, anger, disappointment, in various forms. Maybe there is no unified core feeling of 'fear' or 'joy' that is the same from instance to instance. No matter. Maybe all there is to emotional experience is various sorts of somatic, sensory, and imagery experiences. That doesn't matter either. Think of some occasions on which you have vividly felt what you would call an emotion. Add those to your list of examples of phenomenal consciousness.

Thinking and desiring. Probably you've thought to yourself something like 'what a jerk!' when someone has behaved rudely to you. Probably you've found yourself craving a dessert. Probably you've stopped to try to plan out, deliberately in advance, the best route to the far side of town. Probably you've found yourself wishing that Wonderful Person X would notice and admire you. Presumably not all of our thinking and desiring is phenomenally conscious in the intended sense, but presumably any instances you can now vividly remember or create are or were phenomenally conscious in the intended sense. Add these to your stock of positive examples. Again, it doesn't matter if these experiences aren't clearly differentiated from other types of experience that we've already discussed.

Dream experiences. Although in one sense of 'conscious' we are not conscious when we dream, according to both mainstream scientific psychology and the folk understanding of dreams, dreams are phenomenally conscious — involving sensory or quasi-sensory experience, or maybe instead only imagery, and often some emotional or quasi-emotional component, like dread of the monster who is chasing you.

Other people. Bracketing radical scepticism about other minds, we normally assume that other people also have sensory experiences, imagery, emotional experiences, conscious thoughts and desires, and dreams. Count these, too, among the positive examples.

Negative examples. Not everything going on inside of your body is part of your phenomenal consciousness. You do not, presumably, have phenomenally conscious experience of the growth of your fingernails, or of the absorption of lipids in your intestines, or of the release of growth hormones in your brain — nor do other people experience such things in themselves. Nor is everything that we normally classify as mental part of phenomenal consciousness. Before reading this sentence, you probably had no phenomenal consciousness of your disposition to answer 'twenty-four' when asked 'six times four'. You probably had no phenomenal consciousness of your standing intention to stop for lunch at 11:45. You presumably have no phenomenal consciousness of the structures of very early auditory processing. If a visual display is presented for several milliseconds and then quickly masked, you do not have visual experience of that display (even if it later influences your behaviour). Nor do you have sensory experience of every aspect of what you know to be your immediate environment: no visual experience of the world behind your head, no tactile experience of the smooth surface of your desk that you can see but aren't presently touching. Nor do you have pain experience, presumably, in regions outside your body, nor do you literally experience other people's thoughts and images. We normally think that dreamless sleep involves a complete absence of phenomenal consciousness.

Phenomenal consciousness is the most folk-psychologically obvious thing or feature that the positive examples possess and that the negative examples lack. I do think that there is one very obvious feature that ties together sensory experiences, imagery experiences, emotional experiences, dream experiences, and conscious thoughts and desires. They're all *conscious experiences*. None of the other stuff is experienced (lipid absorption, the tactile smoothness of your desk, etc.). I hope it feels to you like I have belaboured an obvious point. Indeed, my argumentative strategy relies upon this obviousness.

You must not try to be too clever and creative here! Of course you could invent a *new and non-obvious* concept that fits with the examples. You could invent some quus-like feature or 'Cambridge property' like *being conscious and within 30 miles of Earth's surface* or *being referred to in a certain way by Eric Schwitzgebel in this essay*. Or you could pick out some scientifically constructed but folk-psychologically non-obvious feature like accessibility to the 'central workspace' or in-principle-reportability-by-a-certain-type-of-cognitive-mechanism. Or you could pick out a feature of the sort Frankish suggests, like 'quasi-phenomenality' or presence of the disposition to *judge* that one is having wonderful conscious experiences. None of those are the feature I mean. I mean the *obvious* feature, the thing that kind of smacks you in the face when you think about the cases. That one!

Don't try to analyse it yet. Do you have an analysis of 'furniture'? I doubt it. Still, when I talk about 'furniture' you know what I'm talking about and you can sort positive and negative examples pretty well, with some borderline cases. Do the same with phenomenal consciousness. Even you can do this, Keith! Let yourself fall into it. Save the analysis, reduction, and metaphysics for later.

Maybe scientific enquiry and philosophical reflection will reveal all the examples to have some set of functional properties in common, or to be reducible to certain sorts of brain processes, or whatever. That's fine — a variety of unifying features can also be found for 'rectangle' and probably 'furniture'. This doesn't prevent us from defining such terms by example while remaining open-minded and non-commissive about theoretical questions that might be answered in later enquiry.

3. Contentious cases and wonderfulness

Some consciousness researchers think that phenomenal consciousness is possible without attention — for example, that you are constantly phenomenally conscious of the feeling of your feet in your shoes even though you rarely attend to your feet or shoes. Others think consciousness is limited only to what is in attention. Some consciousness researchers think that phenomenal consciousness is exhausted by sensory, imagery, and emotional experiences, while others think that phenomenal consciousness comes in a wider range of uniquely irreducible kinds, possibly including imageless thoughts, an irreducible sense of self, or feelings of agency.

I have avoided committing on these issues by restricting the examples in Section 2 to what I think are likely to be uncontentious cases. I did not, for example, list a peripheral experience of the feeling of your feet in your shoes among the positive examples, nor did I list a non-conscious knowledge of the state of your feet among the negative examples. This leaves open the possibility that there are two or more fairly natural concepts that fit with the positive and negative examples and differ in whether they include or exclude such contentious cases. For example, if phenomenal consciousness substantially outruns attention, both the intended concept of *phenomenal consciousness* and the narrower concept of *phenomenal-consciousness-along-with-attention* adequately match the positive and negative examples.

Similarly, consciousness might or might not always involve some kind of reflective self-knowledge, some awareness of oneself *as* conscious. I intend the concept as initially open on this question, prior to careful introspective and other evidence.

You might find it introspectively compelling that your own stream of phenomenally conscious experience does, or does not, involve constant experience of your feet in your shoes, or reflective self-knowledge, or an irreducible sense of agency. Such confidence is, in my view, often misplaced (Schwitzgebel, 2011). But regardless of whether such confidence is misplaced, the intended concept of *phenomenal consciousness* does not build in, *as a matter of definition*, that consciousness is limited (or not) to what's in attention, or that it includes (or fails to include) phenomena such as an irreducible awareness of oneself as an experiencing subject. If it seems to you that there are two equally obvious concepts here, one of which is definitionally commissive on such contentious matters and another of which leaves such questions open to introspective and other types of evidence, my intended concept is the less commissive one. This is in any case probably the more obvious concept. We can *argue* about whether consciousness outruns attention; it's not normally antecedently stipulated.

It is likewise contentious what sorts of organisms are phenomenally conscious. Do snails, for example, have streams of phenomenally conscious experience? If I touch my finger to a snail's eyestalk, does the snail have visual or tactile phenomenology? If *phenomenal consciousness* meant 'sensory sensitivity' we would have to say yes. If *phenomenal consciousness* meant 'processes reportable via a cognitively sophisticated faculty of introspection' we would have to say no. I intend neither of these concepts, but rather a concept that doesn't settle the question as a straightforward matter of definition — and again I think this is probably the more typical concept to latch onto in any case.

It is this openness in the concept that enables it to meet the wonderfulness condition I introduced at the end of Section 1. One can *wonder* about the relationship between phenomenal consciousness and reportability, wonder about the relationship between phenomenal consciousness and sensory sensitivity, wonder about the relationship between phenomenal consciousness and any particular functional or biological process. One can wonder whether your stream of phenomenal consciousness could survive your bodily death. Maybe a bit of investigation will definitively settle these questions. Wonder doesn't have to be permanent. Wonder is compatible even with the demonstrable mathematical impossibility of some of the epistemically open options: before doing the calculation, one can wonder if and where the equation $y = x^2 - 2x + 2$ crosses the x axis. The 'wonderfulness' condition as I intend it here does not require any kind of insurmountable 'epistemic gap' — only a moment's epistemic breathing space.

I suggest that there is one folk-psychologically obvious concept, perhaps blurry-edged, that fits the positive and negative examples while leaving the contentious examples open and permitting wonder of the intended sort. That's the concept of phenomenal consciousness.

4. Problematic assumptions?

Back to Frankish's dilemma. We get poked by Horn 1 if we commit to anything metaphysically or epistemically dubious in committing to the existence of phenomenal consciousness. We get poked by — or rather, warmly invited to — Horn 2 if we end up with so deflationary a concept of phenomenal consciousness that it ends up just being some 'easy' straightforwardly functional or physical concept.

Frankish offers a nice list of dubious commitments that I agree it would be good not to build into the definition of consciousness. Let me now disavow all such commitments — consistently, I hope, with everything I have written so far. Phenomenally conscious experiences need not be simple, nor ineffable, nor intrinsic, nor private, nor immediately apprehended. They need not have non-physical properties, be inaccessible to third-person science, or be inexplicable in physical terms. My definition by example did not, I believe, commit me on any such questions. My best guess is that all of those claims are false, if intended as universal generalizations about phenomenal consciousness. (However, if some such feature turns out to be present in all of the examples and thus, by virtue of its presence, in some sense indirectly or implicitly 'built into' the definition by example, so be it.

Such indirect commitments to features that are actually universally present shouldn't implausibly inflate our target. This is different, of course, from commitment to features *falsely believed* to be universally present.)

My definition did commit me to a fairly strong claim about folk psychology: that there is a single obvious folk-psychological concept or category that matches the positive and negative examples. But that's a rather different sort of commitment.

I also committed to realism about that concept or category: the folk category is not empty or broken but rather picks out a feature that (most of) the positive examples share and the negative examples presumably lack. If the target examples had nothing important in common and were only a hodgepodge, this assumption would be violated. This is a substantive commitment, but not a dubious one I hope. (However, if the putative negative examples failed to be negative, as in some versions of panpsychism, we might still be able to salvage the concept, by targeting the feature that the positive examples have and that the negative examples are *falsely assumed* to lack.)

The wonderfulness condition involves a mild epistemic commitment in the *neighbourhood* of non-physicality or non-reducibility. The wonderfulness condition commits to its being not straightforwardly obvious as a matter of definition what the relationship is between phenomenal consciousness and cognitive functional or physical processes. This commitment is quite compatible with the view that a clever *a priori* or empirical argument could someday show, perhaps even has already shown, that phenomenal consciousness is reducible to or identical to something functional or physical.

Frankish's quasi-phenomenality, characterized in terms of our dispositions to make phenomenal judgments, does not appear to meet the wonderfulness condition (see also Frankish's, 2012, 'zero qualia'). It does not leave open, even for a moment, the question of whether phenomenal consciousness might be present even in the absence of a certain cognitive functional feature: the disposition to make phenomenal judgments. I do think that question is open, at least for a moment — and probably for much more than a moment. I wonder, for example, whether snails might be conscious despite (presumably) their not being disposed to reach phenomenal judgments about their experience. I wonder whether we might have fleeting, unattended conscious experiences even if we are not disposed to reach judgments about them. I even wonder whether group entities like the United States might possess phenomenal consciousness at a group level, despite (presumably) no tendency to judge that they are doing so (though I doubt many people will join me in wondering about this).

After being invited to consider the positive and negative examples, someone might say, 'I'm not sure I understand. What *exactly* do you mean by phenomenal consciousness?' At this point, it is tempting to clarify by making some epistemic or metaphysical commitments — whatever commitments seem plausible to you. You might say, 'those events with which we are most directly and infallibly acquainted' or 'the kinds of properties that can't be reduced to physical or functional role'. Please don't! Or at least, don't build these commitments into the definition. Such commitments risk introducing doubt or confusion in people who aren't sure they accept such commitments. Maybe it's okay to say, 'that about which it has often been *believed* we have direct, infallible access and *believed* to be irreducible to the physical'. But let the examples do the work.

Here's a comparison: you are trying to teach someone the concept 'pink'. Maybe her native language doesn't have a corresponding term (as we don't have a widely used term for pale green). You have shown her a wide range of pink things (a pink pen, a pink light source, a pink shirt, pictures and photos with various shades of pink in various natural contexts); you've verbally referenced some famously pink things such as cherry blossoms and ham; you've shown her some non-pink things as negative examples (medium reds, pale blues, oranges, etc.). It would be odd for her to ask, 'so do you mean this-shade-and-mentioned-by-you?' or 'must "pink" things be less than six miles wide?' It would be odd for her to insist that you provide an analysis of the metaphysics of pink before she accepts it as a workable concept. You might be open about the metaphysics of pink. It might be helpful to point, non-committally, to what some people have said ('well, some people think of pink as a reflectance property of physical objects'). But lean on the examples. If she's not colour-blind, and not perverse, there's something obvious that the positive instances share, which the negative examples lack, which normal people will naturally latch onto well enough, if they don't try too hard to be creative or insist on an analysis first, and if you don't confuse things by introducing dubious theses: they're all *pink*. This is a perfectly good way to teach someone the concept *pink*, well enough that she can confidently affirm that pink things exist (perhaps feeling baffled how anyone could deny it), sorting future positive and negative examples in more or less the consensus way, except perhaps in borderline cases (e.g. near-red) and contentious cases (e.g. someone's briefly glimpsed socks). My view is that the concept of *phenomenal consciousness* can be approached in the same manner.

I want, and I think we can reasonably have, and I think the most natural understanding of 'consciousness' already gives us, room to wonder about certain things. We needn't commit straight away to either a reductionist picture on which everything is physical stuff, entirely mundane, or to what Frankish calls a 'radical realist' picture on which consciousness somehow transcends the physical. If I had to bet, I'd bet on the mundane, but I don't want to build it right into my conceptualization of consciousness. I want as innocent a concept as I can manage, which leaves the possibilities epistemically open.[1]

References

Block, N. (1995/2007) On a confusion about a function of consciousness, in *Consciousness, Function, and Representation*, Cambridge, MA: MIT Press.

Chalmers, D.J. (1995) Facing up to the problem of consciousness, *Journal of Consciousness Studies*, **2** (3), pp. 200–219.

Chalmers, D.J. (1996) *The Conscious Mind*, New York: Oxford University Press.

Frankish, K. (2012) Quining diet qualia, *Consciousness & Cognition*, **21**, pp. 667–676.

Schwitzgebel, E. (2011) *Perplexities of Consciousness*, Cambridge, MA: MIT Press.

Searle, J.R. (1992) *The Rediscovery of the Mind*, Cambridge, MA: MIT Press.

Siewert, C. (1998) *The Significance of Consciousness*, Princeton, NJ: Princeton University Press.

---

[1] For helpful discussion, thanks to Keith Frankish, Pauline Price, and commenters on my related blog post at the Splintered Mind.

James Tartaglia

What is at Stake in Illusionism?

Abstract: *I endorse the central message of Keith Frankish's 'Illusionism as a Theory of Consciousness': if physicalism is true, phenomenal consciousness must be an illusion. Attempts to find an intermediate position between physicalist illusionism and the rejection of physicalism are untenable. Unlike Frankish, however, I reject physicalism, while still endorsing illusionism. My misgivings about physicalist illusionism are that it removes any rational basis from our judgment inclinations concerning consciousness, undermines the epistemic basis required to explain the genesis of our physical conception of the world, and leads to a widespread scepticism about the basis of philosophical reflection. I endorse the core of physicalist intuition, but not its metaphysic, and sketch my alternative illusionism, which resists physicalism's merging of philosophy with science without thereby impinging on science. I conclude that physicalism is fostered by inattention to metaphilosophy and threatens philosophy's distinctive voice; but that illusionism itself is an important insight.*

Keywords: phenomenal consciousness; physicalism; illusionism; eliminative materialism; metaphilosophy; transcendence; identity theory; phenomenal concept strategy.

1. Towards another kind of illusionism

As someone with strong illusionist sympathies, I found myself agreeing with so much of what Keith Frankish says in this refreshingly clear and insightful essay that I formed the hope that it will become a landmark in the philosophy of consciousness from which future

Correspondence:
James Tartaglia, Keele University, Keele, Staffordshire.
Email: j.tartaglia@keele.ac.uk

discussions take their lead. For with the issues set up this well, the debate might be able to move away from the features which have characterized it in the last twenty-plus years: lack of vision combined with tiny dialectical moves, needless proliferation of terminologies, and the tedious reciting and consequent dissection of a few simple thought experiments. With this paper as a new starting point, the debate might take a fresh turn and become more interesting. I think it deserves to acquire this status, because the central case it makes is of crucial philosophical importance. It is that if physicalism is true, then consciousness must be a kind of illusion. All the murky, intermediate positions trying to hold on to both consciousness (conceived in the ordinary, phenomenal way) and physicalism (conceived in the ordinary, non-revisionary way) are untenable. The philosophers at the opposite poles, namely anti-physicalists and illusionist physicalists, have been saying this for years. They were right. Attractive as it might seem, there is no stable middle ground on this issue. You cannot have your cake and eat it: consciousness (as ordinarily conceived) or physicalism (ditto) has to go.

Frankish makes this case extremely well, but it is worth reminding ourselves of the *prima facie* incompatibility from which it derives, lest we become tempted by the thought that a subtle philosophical distinction might save the day. Phenomenal properties are conceived in direct opposition to physical properties, namely as subjective properties typically caused by objective, physical properties. Thus phenomenal-red, for instance, is subjective in that, unlike physical properties, it exists only from the perspective of the person experiencing it, and to get a grasp on its notion requires us to contrast the redness a flower may possess with the redness the flower causes within us. Phenomenal-red can only be introspected; red can only be perceived. The former is only there for me; the latter for everyone. In short, to understand the notion of phenomenal-red, you have to contrast it with red; that is how we all get a grip on the notion when we start studying philosophy. So the starting point for would-be intermediates is rather like that of trying to show that our conceptions of a square and of a circle are really just two conceptions of the same thing.

If we are conceiving the same thing in directly opposed ways, then one of these conceptions must be wrong. But in that case, since the physical conception is the one conventional physicalists trust to tell us how the world is, the intentional object conceived by the phenomenal concept cannot really be there. There cannot be anything in the world that fits the phenomenal conception, if it is directly opposed to another that provides an accurate conception of the same thing. So if the phenomenal concept is a concept of a brain state, it must be a radical misconception of it; we must be misconceiving the brain state beyond all recognition, in fact. We are thinking of a brain state as a subjective experiential array, but that is not what it is at all. Consequently, the array must be an illusion, even if thinking about it somehow allows us to think about real brain states.

Like Frankish, then, it seems perfectly clear to me that we cannot pretend our phenomenal conception of consciousness is OK if we think conventional physicalism is true. Our phenomenal concepts cannot be getting it right if our physical ones are, so if we are to preserve the *prima facie* situation, we must either accept that they are presenting us with illusions, abandon physicalism, or move into the dubious territory of making predictions about the future of physics. The only option for philosophers who want to both have and eat cake, then, is to try to cast doubt on the *prima facie* conceptual incompatibility. But since it concerns how we are *conceiving* of the world, this project is doomed from the outset. For this is a factual matter; we cannot make our ordinary conception of consciousness something it is not. The insurmountable obstacle to the would-be intermediate, then, is that we *do* think about consciousness in a way which is problematic for physicalism; which is something they concede by entering the debate. They could resolve the problem by arguing that this conception is wrong, as the illusionist does, but there is no room to argue that we have the conception we do *and* that it is unproblematic for physicalism.

This is the standard approach nonetheless. Thus the mainstream of physicalism since the 1990s, inspired by Loar (1990), has tried to find a middle way by claiming that phenomenal concepts are empty and non-committal. The basic idea is that the concept of phenomenal-red, for instance, does not really tell us anything about its referent, but rather just directly picks it out — 'directly' in the sense that it does not connote any features of its referent, in the way the concept of Phosphorus connotes a morning appearance. This move is supposed to vindicate our ordinary conception of consciousness on the grounds that the conception of phenomenal-red is not telling us anything incompatible with physicalism — because it is not telling us anything about its referent at all, just referring to it. Thus the *prima facie* incompatibility is supposed to be removed in a manner which leaves phenomenal concepts intact.

The root problem with this proposal remained hidden for many years, due mainly to the wide variety of subtly different forms in which it was defended (it took about two decades to even acquire a standard name: 'The Phenomenal Concept Strategy'). But now the dust has settled, I think it is plain that its central claim about the nature of phenomenal concepts is both unsupported and patently false. Phenomenal concepts are rich and theoretically committal. If I look at a flower and think about what I am aware of as phenomenal-red, then I think my attention is focused upon an inner experiential array of phenomenal-colour, quite unlike anything to be found either on the surface of the petals or inside my brain. So as a matter of fact, phenomenal concepts are not as the phenomenal concept strategy describes them. I have previously laboured this point (Tartaglia, 2013), which Frankish gets across nicely in two paragraphs (this issue, pp. 25–6).

The phenomenal concept strategists did not argue for their 'empty' reading of phenomenal concepts; rather they defended it as a neglected theoretical alternative capable of solving the problem of consciousness (Tartaglia, 2013). If our concepts were as they say they are, however, the problem would never have arisen. So since these philosophers must originally have been working with the ordinary conception, their 'empty' reading must be revisionary. The phenomenal concept strategists forgot that a person's *conception* of something is how that something seems to be to the person. They must have forgotten, since they argue, in effect, that although it seems to us that phenomenal states are private presentations with no place in the physical world, they also seem to be nebulous occurrences whose nature completely eludes us. This is obviously incoherent unless a temporal distinction is introduced, thereby revealing the proposal to be the overtly revisionary one of abandoning our original conception in favour of a new one. But then they cannot claim to have found a way of maintaining the *prima facie* situation.

Construed as a proposal to amend our phenomenal thinking, the phenomenal concept strategy asks the impossible of us. I cannot think of my experiences as presenting themselves as I-know-not-whats, because they present themselves as very familiar subjective states, the likes of which I have never encountered in the objective world (and cannot imagine how I could). Like Frankish, I take these presentations to be illusory, but I cannot eradicate my conception of them in favour of something more neutral; only build on it with the extra thought that they are illusory. To remove the *prima facie* conceptual incompatibility, however, this is what the phenomenal concept strategy asked of us. It took its lead from the identity theorists of the 1950s, with their original proposal that our experiential concepts are 'topic-neutral', such that we think of experiences as simply 'something going on' in specifiable circumstances. Place (1956) and Smart (1959) showed considerably more insight, however, because they recognized that this would require phenomenal thinking to be renounced.

That we do not conceive of our experiences in this anaemic manner did not present the same problem for the identity theory as it does for the phenomenal concept strategy. For this broadly functionalist conception is at least part of what we intend, and Place and Smart were quite clear that they thought the other, phenomenal part was untenable; the phenomenal concept strategists, by contrast, claimed to have preserved it. What the identity theorists were not clear about, quite excusably given that they were pioneers, was the implication that our ordinary conception of consciousness is of an illusion. They were proposing a new way of thinking about conscious states that was compatible with physicalism, and do not seem to have thought much about the old one. But it did not go away, and the intuitions it continued to engender kept physicalism mired in controversy. Things took a turn for the worse when physicalists who had been persuaded by Nagel (1974) and Jackson (1982) that the phenomenal conception was legitimate tried to combine it with physicalism in order to find that elusive middle way. With this *desideratum* set, then once Kripke's (1972) causal theory of reference entered the mix, we found ourselves landed with the explanatory gap and the phenomenal concept strategy. Thus the original illusionist thrust of physicalism did not last long, being preserved only by the likes of Rorty (1982) and Dennett (1991). It now seems to be regaining focus, however, and this I welcome; hence my hope for the fortunes of Frankish's paper.

Now given what I have said so far, and the fact that not so long ago I published a paper defending a physicalist form of illusionism along very similar lines to Frankish (Tartaglia, 2013), you will probably not be expecting what comes next (unless you read the abstract). For I now reject physicalism, though not illusionism. As I have said, I think the only stable options in this debate are physicalist illusionism or a rejection of physicalism. I opt for the latter, and yet I am still an illusionist.

You might well find this combination puzzling, if you think of illusionism as simply a consequence of physicalism. I do not.

Evidently, then, I do not reject physicalism because I think that illusionism is obviously crazy, as many anti-physicalists do. And I am not about to defend dualism, panpsychism, or any of the other metaphysical positions familiar in this debate; my views on such positions are similar to those of most conventional physicalists. Neither am I going to reject physicalism on the grounds that it cannot be rigorously formulated; such considerations do not move me. Rather I shall be taking things up to the metaphilosophical plain, which is something I learnt to do from Rorty, one of the great illusionists.[1]

2. The problems of physicalist illusionism

Any credible metaphysic must be compatible with the manifest situation which we all, as conscious beings, find ourselves in. So it must be at least possible that what the metaphysic says about consciousness explains why, for instance, I seem to be here in my office on a certain Monday morning, with a point of view that takes in my computer screen (as viewed from a certain location and with certain eyes), my awareness of my thoughts in writing, and sensations such as the coffee-taste lingering in my mouth. If the metaphysic would rule out this being my manifest situation, were it true, then it is not true. The manifest situation will not go away no matter how we metaphysically portray it, as Frankish realizes (p. 20); if our metaphysical understanding of what is going on is incompatible with it, then we are simply confused. That is why 'illusionism' is a better name than 'eliminativism'; for we cannot eliminate the manifest situation.

This, I think, is the core wisdom to Chalmers' (1996) injunction to 'take consciousness seriously'. Stripped of any further connotations, this might just seem obvious. And indeed it is; but plenty of physicalists have overlooked it by focusing on the details of the metaphysic (physical vs. functional, for instance), without concerning themselves with how *that* could be true given our manifest situation. Without an attempt at reconciliation, however, consciousness is not being taken seriously. Illusionists like Frankish and Dennett take it seriously. Dennett demonstrates this by sandwiching his account of consciousness between two renditions of a vivid example of the kind of manifest situation that needs to be explained — by the time of the

---

[1] Arguably, he was its originator. See Tartaglia (2016a) for my take on Rorty's illusionism; and for an unusual exposition of his position (written as if by him), see Tartaglia (2016b).

second, he takes himself to have done the job (1991, pp. 26–7 & p. 407).[2] I do not think any philosophers really doubt the manifest situation, although they are sometimes disingenuously accused of doing so; as when Searle invites us to pinch ourselves as a preface to his critique of Dennett (Searle, 1997, pp. 97ff.). Such talk is best construed as saying, elliptically and with frustration, that the metaphysic is not compatible with the manifest situation, not that the theorist is unaware of that situation (or somehow does not share it). The problem of philosophers not taking consciousness seriously is not that of denying the manifest situation, but rather of not providing anything meant to explain it.

Frankish seeks to explain it with intentionality, as I did when I was a physicalist. The idea is that the phenomenal arrays we are apparently presented with are objects of intentional acts of mind. As Brentano emphasized, such objects need not be real: we can think about real people, but also fictional ones like Sherlock Holmes. The manifest situation, then, is supposed to be explained by the fact that human beings have intentional states about fictions, i.e. phenomenal states. This, Frankish suggests, explains their apparent causal potency; they 'move us in the same way that ideas, stories, theories, and memes do, by figuring as the objects of our intentional states' (p. 27).

The initial difficulty this proposal faces is that thinking about fictional objects like Holmes *feels* like something; this is even more obvious in the case of imagining them.[3] We do not typically find the Holmes case puzzling because we do not bring the concrete phenomenological presence into dispute when we claim that it is not indicative of a real man. Illusionists must, however, so the parallel with thinking about fiction will not take them far before they must reconstrue it. They must claim that any apparent presence is itself a merely intentional object, not a real one. But then, how is it that it *feels* like something? As we understood this before, conscious awareness of fictional objects was facilitated by phenomenological presences. So if the presences are themselves to be reduced to intentional objects, we need a new account of the appearance of a concrete presence which occurs when we think about illusory objects; it can no longer be a real presence.

---

[2] For a critique of Dennett based around this example, see Tartaglia (2016c, pp. 92–5).

[3] The illusionist should agree: if they are trying to account for our manifest situation, they should not concede ordinary terms like 'feels' to phenomenal realists.

The account both Frankish (pp. 29–34) and Dennett (1991, pp. 364ff.) provide is that the appearance of a concrete presence is a matter of our judging or representing that there is one. And we certainly do judge this, even if we then correct ourselves. So the issue is now recast in terms of the basis of this judgment inclination; whether we judge this way, and hence it seems this way, because of a real presence or not. The point the physicalist illusionist overlooks at this juncture, however, is that if we opt for 'not', as they suggest, then the judgment becomes evidentially baseless. In essence, the view is that when I seem to see a vivid patch of red on closing my eyes — that is, when I am inclined to judge there is such a presence, even if my illusionist commitments make me subsequently refrain — then there is no rational basis for my inclination whatsoever. My judgment inclination is rationally inexplicable from my own personal perspective, even if a causal explanation can be given for how my central nervous system is affecting my speech centre, say. This must be the case because, as physicalists of all persuasions continually emphasize, there is no evidence for the existence of phenomenal consciousness in the physical world. Nothing like that vivid red patch is to be found in the brain or anywhere else; which is why astute physicalists think it must be an illusion. But then, whatever physicalist illusionists think my judgment inclination is actually sensitive to, it cannot be a reasonable interpretation of the data.

This is the consequence of the physicalist illusionist's substitution of bare judgment for phenomenology as an account of seeming. It is unavoidable, because they need this account of seeming for the intentional object account of phenomenology which makes their account explanatory; and they need the claim that there is no evidence for the existence of phenomenal presences for their physicalism. As I have said elsewhere (Tartaglia, 2016c, p. 94), the result is an account according to which our irrepressible urge to make judgments about the phenomenal character of our experiences is somewhat akin to suffering from Tourette's syndrome — in that something in our wiring must be making it practically impossible for us to resist sincerely uttering sentences like 'there is a vivid red patch', even though we have no evidence for the presence of such a patch, and even though we may be sold-up physicalist illusionists.

Two further consequences of making this move are as follows — the second will begin to take us into the metaphilosophical heart of the matter. The first is that if these judgment inclinations are rooted in an illusion, then our confidence in physicalism must immediately be shaken. This is because we drew up our physical conception of the world on the basis of (phenomenal) conscious experience.[4] You could deny this outright by claiming that we would have come up with the same conception even without conscious states; that consciousness is epistemically epiphenomenal. If a claim this implausible were required by illusionism, however, we could dismiss it out of hand. But if we concede, as we must, that consciousness had a role in shaping our conception of the world, the illusionist claim that conscious states inspire us to believe in non-entities immediately places that conception in doubt. If a book central to a thousand years of Egyptology was discovered to be completely unreliable, our understanding of Ancient Egypt would be thrown into disarray. You do not need the foundationalist conviction that consciousness is our ultimate sourcebook to see that if judgments about conscious experiences are completely unreliable, then given how thoroughly intertwined they are with the genesis of our physical conception of the world, our confidence in that conception must be profoundly shaken. And our confidence in illusionism must be too. For how, on the basis of an illusion, could we discover true reality, and hence be confident that our evidence was indeed illusionary?

Since representations of phenomenal presences are misrepresentations, according to the illusionist, what they accurately represent must be brain states or distal objects. Frankish considers both options (p. 19), saying that consciousness might be an introspective illusion (brain states) or transparent (distal objects). The idea that they represent brain states holds particular appeal for physicalists because, on their account, that is what the representations ultimately are; hence they are representing themselves, which promises an account of self-consciousness. But then the misrepresentation could hardly be more radical; for brain states are nothing like vivid red patches. This is clearly quite unlike paradigm cases of misrepresentation where a representation which has its content fixed by sheep, say, is inadvertently triggered by the presence of goats; or Frankish's

---

[4] Frankish only says that the phenomenal aspect of experience is illusory, not experience itself, which can be understood functionally. I made the same distinction in Tartaglia (2013), following the lead of the original identity theorists. However we are only aware of our experiences, such that they can enter into our reasoning about what caused them, what their function is, etc. because of what phenomenal realists understand as their phenomenal aspects. Since the distinction will not affect my case, I will hereafter omit it.

example of misrepresenting a flat hologram as a three-dimensional object (p. 36).

The situation is little changed if we think of conscious presences as misrepresentations of distal objects. On the face of it, this would allow us to invoke the primary/secondary quality distinction to find an element of accurate representation. The illusory appearance of an apple does at least seem to track its shape and size, and that starting point might allow us to determine that its illusory phenomenal-colour tracks certain types of physical surface. But this is to forget that the shape we discern in the experience is itself illusory (Tartaglia, 2016c, p. 110). It is the shape of an illusory phenomenal presence; or, if we insist upon transparency,[5] it is the shape outlined by an illusory presence projected onto the object. Since my conception of this shape is a misrepresentation, then, I am in no position to assert any kind of isomorphism between the experience and apple. I have no reason to think my mistake can tell me anything about the shape of the apple, and hence no basis to draw up the functional correlations between conscious states and distal conditions which physicalism relies upon.

Frankish looks hard for veridical elements to the illusion; perhaps phenomenal illusions and brain activity might be thought to share 'intensity' (p. 17), he suggests. But if I am wrong about the intensity of the experience, since phenomenal intensity is an illusion, then I have no basis to infer that it is indicative of a real feature. Since I can only cherry-pick the illusion for veridical elements once physicalism tells me what those veridical elements might be, this tactic cannot render the *genesis* of its conception of reality plausible; instead it is a tactic which simply dilutes the metaphysical insight of illusionism for the sole purpose of trying to preserve the physicalism it epistemically undermines.

Perhaps the problem is alleviated if we do not think of representation as causal covariation, but rather in terms of the functional-role semantics Frankish prefers (p. 36). I think this only makes matters worse. For then our representations have their content fixed by abstract brain functions, but misrepresent those functions as concrete presences; the misrepresentation must now cross the metaphysical divide between abstract and concrete, in addition to everything else it gets wrong. Nevertheless, supposing the magic trick is performed, we could then represent the function of the illusory phenomenal presences

---

[5] I would not; see Tartaglia (2016c, pp. 113–4).

through how they relate to each other; that is, represent the functions of misrepresentations of brain functions (e.g. the functions of illusory red patches). But then to glean anything about the external world from that, we must correlate these functions with external conditions (e.g. red surfaces), and the only way to do that is to delve into the illusory content of the misrepresentations again (their phenomenal redness), thereby taking us back to the original problem.

By rendering consciousness illusory for the sake of our physical conception of the world, the physicalist forgets where that conception came from. With the physical conception in place, physicalist illusionism makes sense. But human beings built up that conception by working out what their conscious experiences were telling them about the world. If fictions were integral to our reasoning, we should never have believed the physical conception that tells us they are fictions. Frankish suggests that consciousness is an illusion hardwired into our psychology by evolution (p. 37). But this view is inherently unstable when presented against the backdrop of physicalism. For why should we trust an evolutionary illusion to lead us to the true nature of reality? As Dennett says, 'If some creature's life depended on lumping together the moon, blue cheese, and bicycles, you can be pretty sure that Mother Nature would find a way for it to "see" these as "intuitively just the same kind of thing"' (Dennett, 1991, p. 381). Evolution cares about survival, not getting things right. Perhaps getting it right explains how we have survived, but this is hard to maintain when pressing the case that our starting point was an illusion; for conscious experiences guided us to evolutionary theory.

This leads onto the second consequence, which is that physicalist illusionism renders our manifest situation illusory in a wholesale manner that goes far beyond debates about consciousness. For consciousness provides the basis of the intuitive way we have of thinking about our lives and world. Many aspects of this have been central to philosophical debates for millennia — from freedom to personal identity to conscience — but a particularly vivid one in this context is time. My experience tells me that this particular point in time is where I am in my life at the moment, and hence has a special significance (soon to be lost). But physicalist illusionism tells me this is no basis for a serious judgment. I should only believe the results of collective, scientific rationality; for physicalism holds that science must determine our metaphysics. I then find that contemporary physics strongly suggests a 'block universe' conception of reality, according to which my personal 'now' has no ontological significance. So I am not currently aware of even illusory phenomenal presences in any substantive sense. Rather, I am reporting an instantaneous state of four-dimensional reality; a projection or three-dimensional 'time-slice' on a par with any other projection of the four-dimensional reality that constitutes my life. In a very real sense, then, I am not even *here*; or better, I am everywhere that, while alive, I ever sincerely want to call 'here'. Of course, some philosophers dispute the block conception, but since their starting point is how consciousness presents our manifest situation, physicalist illusionism denies them any good grounds for doing so.[6]

Does this mean that physicalist illusionism is incompatible with our manifest situation, and consequently false? I do not think we can answer this without determining whether physicalism is true. Frankish accepts that we have 'introspective subjectivity', which need not amount to 'intrinsic subjectivity' (p. 31), going on to say that, for a (physicalist) representational theory of mind, the difference between a non-veridical introspective representation of a feeling and a veridical one is subjectively 'no difference at all' (p. 34). But without presupposing physicalism, I see no way of determining whether I could sincerely make judgments about my manifest situation in the absence of real phenomenal presence. Even if we made a machine that enacted the perfect illusionist programme, I could not step inside its introspective subjectivity to see if its judgments were sincere. I could not *become* the machine, only implement functionally identical representations in my own body; that of a conscious being.[7] If physicalism is true, then I agree with Frankish that the machine's judgments would have to be sincere; but if it is not, they might not be. The real crux of the issue, then, is whether physicalism is true. Since illusionism undermines physicalism's evidential basis, I think we already have good reason to suspect that it is not. But with an alternative on the table capable of satisfying the core of physicalist intuition, I think we can do better.

---

[6] See Tartaglia (2016c, chapter 6).

[7] If on interfacing with the machine I were to lose consciousness, scientists would simply adjust the program. When they got it 'right', this might simply indicate their ability to alter an extant consciousness, rather than manufacture it in a machine.

3. The nature of the illusion

The core of physicalist intuition, it seems to me, is that natural science provides our best vocabulary for describing what actually exists. So when philosophers put forward consciousness as an obstacle to scientific explanation, this immediately seems dubious; given that philosophy has only *a priori* reflection, and science has equipment like the Large Hadron Collider! It sounds like either a throwback to the bad old days when philosophers tried to practise natural science *a priori*, as in idealist works of *Naturphilosophie*, or else just a badly informed defeatism about the limits of science, destined to be disproved by the next scientific breakthrough, if indeed one is even needed in this case. Speculation about a non-physical reality which interacts with or supplements the physical one seems only slightly more outdated than speculation about a psychical 'inside' to reality and predictions of a currently unimaginable revolution in physics. For, if science is to be supplemented, this will be done by scientists for scientific reasons; philosophers are kidding themselves if they think their reflections will lead the way. I agree with all of that. And once you have that core of intuition, the only place to go is physicalism, right? Wrong.

Physicalism is a philosophical position, according to which our metaphysical conception of reality is to be dictated by physical science. If, like Quine (1975), you do not recognize a distinction between philosophy and science, then this will not strike you as a significant statement, and the core of physicalist intuition will immediately lead you to physicalism. If you do, however, then you will not think that the possibility of science providing a complete objective description of reality automatically settles the nature of the metaphysic we should adopt; because such a description might not capture the independent nature of reality, which is what metaphysics is interested in. To pursue this possibility without abandoning the core of physicalist intuition, you must ensure that your account of this independent nature does not interfere with science's objective picture by requiring anything of it. It must remain a purely philosophical matter, to which science can remain indifferent. This is a possibility which may hold appeal if physicalist illusionism strikes you as having problematic and unattractive consequences. And here is a sketch of how it might pan out, which will in turn shed light on what might be meant by a 'purely philosophical matter'.

Suppose you are currently dreaming. Granted, real dreams about reading philosophy papers may never happen (I hope not), and if they did, they would not have the phenomenological detail and lucidity of your current experience. But Descartes was surely right that you *could* be dreaming right now, even if we have to support the 'could' with science fiction scenarios about brain manipulation. Now suppose that during this (super-)dream, you start to think about the problem of consciousness. You would no doubt find this just as puzzling as in waking life. If you are a physicalist, you would think about both your conscious experiences and the brain states in your head, and wonder how to make sense of the claim that they are the same thing — the impossibility of reconciling these directly opposed conceptions might lead you to illusionism. But note that the supposition that you are dreaming alters the significance of this thought process. For, when thinking about the brain states in your head, you would naturally think of the head within the dream; the one with eyes focused upon these words. But of course, if it is a dream, then there is no prospect of identifying your experiences with *those* brain states, for they are not real ones. They are no more real than anything else in the dream. Rather, if your experiences are to be identified with any brain states, then it must be with the ones inside your real, sleeping head; the one resting on the pillow of your bed. For whatever reality there is to your experiences is not to be found within the dream.

From the experiential perspective of the dream, the world in which you are asleep in bed is a transcendent context — it transcends the space and time of the dream. Now you might think, in line with a venerable philosophical tradition according to which the mind is temporal but not spatial, that the real world transcends only the space, not the time, of the dream. You might think this on the grounds that, although the space of the dream is illusory, it nevertheless takes place in real time; that the events of the dream could be synchronized with the clock above your sleeping head, for instance. However, this thought is confused, as I think must be recognized if we take illusionism seriously — illusionism without the physicalism, that is.

Whether or not this is a dream, the phenomenal presence of the screen I am looking at has a top and bottom. This spatiality is something that could only be found in the objective world, where the phenomenal presence does not belong; so the spatiality is illusory. But if this is a dream, then in order to think of the presence as at least causally connected with brain states, as I must in order to make sense of the possibility of altering it by causally affecting regions of objective space, I must misrepresent the presence as an occupant of the objective world. But then, it is equally true that to conceive the presence as in causal communion with my objective brain, I must think of it as within objective time — as in sync with the clock above my sleeping head. So given that this misrepresentational projection into the objective world gives us no reason to think experiences really possess spatiality, and that dream experiences take place in a unified space and time, I see no more reason to trust dream temporality than spatiality when it leaves the context of the dream. The dream-presences simply do not belong to the objective world; the latter transcends both the space and time of the dream.

If this is a dream, then, any reality there is to my experiences belongs to a context that transcends the dream. I must project them into the objective world in order to explanatorily interweave them with that world, given what I know about it, but in doing so I misrepresent them as really having the spatiality and temporality which they seem to possess in the dream. Now if we take dream experience as indicative of how experience in general works, namely by enclosing us in an illusory experiential world from the perspective of which the independent reality of those experiences could only be found within a transcendent context, then we are ready to form what I call the 'Transcendent Hypothesis' (Tartaglia, 2016c, pp. 101–21). For then we will entertain the thought that consciousness in waking life works the same way. If it does, then the independent reality of our experiences is not to be found in the objective world described by science; flawlessly, at least in principle, and in no need of supplementation on the basis of philosophical speculation. Rather, it belongs to a context of existence which transcends the objective world.[8]

In that case, there is nothing we can cogently say about this independent existence, except that (1) it is transcendent (it is not accurately characterized as belonging to the world of objective space and time, though its existence is responsible for our finding this world[9]), and that (2) it exists. We can say nothing substantive about independent existence; but by misrepresenting it, and thereby thinking

---

[8] Thus the objective world transcends dreams, and independent reality transcends the objective world; independent reality transcends dreams too, but at one step removed. This extra step is required to account for the existence of dream experience, since dreams are immediately transcended by the objective world, where, as all illusionists realize, no kind of (phenomenal) experience can reside.

[9] I work through the details of this in Tartaglia (2016c, pp. 101–67).

of experiences as the kind of things that might be systematically correlated with objective states of affairs, we have arrived at our objective picture of the world in all of its astonishing detail.[10] Freed of physicalism, it is no longer puzzling that we should form objective conceptions of individual experiences to facilitate causal explanations involving the impact of the world on these experiences and *vice versa*, and that once the picture was complete it had no place for our starting point. For a completely objective picture is centreless. This is only puzzling if we try to eliminate the starting point by making it a profoundly misleading revelation of objective reality's independent nature; for then we must engage in desperate *ex post facto* cherry-picking for the elements that might have been revelatory. But the starting point is not eliminated if it is transcendent; the illusion was that it had objective features. Our objective conception of the world is the end product of working out the implications of these supposed features. All of them, for we learnt about light and surfaces from so-called 'phenomenal-colour', and science will ultimately correlate all features of experience with conditions in the outside world, our brains, or both; despite the fact that experiences are absent from the world they are correlated with.

The fact that our objective picture of reality cannot describe independent existence, given the nature of consciousness, is a purely philosophical concern which connects with the core subject matter of the discipline ever since Plato's invocation of transcendence (Tartaglia, 2016c, pp. 61–81; 2016d). It is of no concern to the science of consciousness, which can continue to map the correlations between experiences (misrepresented as objective and reported accordingly) and physical conditions, without bothering itself with the philosophical implications of using the language of causation, identity, or misrepresentation. It can use whatever language is most convenient (surely causation), and ignore the notion that it is 'missing out' something — a notion pressed so successfully by some philosophers that they have spread their own confusion about the nature of their

---

[10] This includes our notion of representation; see Tartaglia (2016c, pp. 147–67). This is not idealism, transcendental or otherwise. Neither is it phenomenal realism, since I am illusionist enough to think that our understanding of independent reality as consciousness is a misrepresentation. However *what* we are misrepresenting (as conscious states in causal communion with an objective world) is certainly real, which is why this account cannot be recast in fictionalist terms acceptable to a physicalist (thanks to Frankish for this suggestion).

discipline into science, such that many scientists are convinced they have a scientific challenge to overcome, or else, as is often the case, that they must denigrate philosophy for suggesting it has not already been overcome. For the only thing being missed out is the metaphysics, and there is a well-established discipline to deal with that. Science aims to perfect our objective representation of the world. But reflecting on the status of this representation within the context of transcendence is a philosophical concern.

Physicalism gained its ascendency in the twentieth century, when the achievements of science became so hard to ignore that it came to seem that any serious investigation of the nature of reality must be scientific. Philosophy wanted to associate itself with science rather than religion, with which science was increasingly at odds, and traditional metaphysics seemed to be, and often was, on the side of religion. But religions make specific claims about the nature of the transcendent context; most typically that it is a meaningful one capable of vindicating human strivings. Such claims are necessarily baseless according to the transcendent hypothesis, which suggests only that the evolution of consciousness, as we conceive it objectively, enclosed us within an illusory phenomenal world which, when thus conceived in objective terms, allowed us to make maximal, objective sense of reality, as well as grasp philosophically that this sense could not capture reality's independent nature.

I think that what led philosophers to embrace physicalism, and the gut-instinct which sustains sympathy with it to this day, is a desire to align philosophy with science against religion and the supernatural. Much-touted motivations such as causal closure, by contrast, only seem powerful against the backdrop of Cartesianism; for closure rules out practically no traditional metaphysical positions except interactionist dualism. Once physicalism had been embraced, due to a lack of historically-informed metaphilosophical self-consciousness which thereafter steadily increased, the remit of philosophy was radically reduced; for the core of metaphysics had been handed over to science. Philosophers began to think that claims not made on the basis of science lacked 'conservatism', as Frankish puts it (p. 24), thereby forgetting to conserve the insights of their own tradition. Philosophy continued to thrive nonetheless, but only because, outside of philosophy of mind, little heed was paid to physicalism and its illusionist consequences, such that philosophers felt able to continue using the manifest situation as their starting point; while inside philosophy of mind, opposition to physicalism provided a continual feed of tasks for physicalists to respond to.

If physicalism were to win the day, philosophy would shrink to a point, and those who have been calling for its replacement by science ever since Comte would finally have their way. This is immediately clear in the case of philosophy of mind. For if it became generally accepted that science tells us what the mind *is*, there would only be so much interest to expend in the question of how best to formulate that insight; remove all the bats, black-and-white scientists, zombies, and Chinese rooms from this area of philosophy, plus their predecessors in the form of misgivings about behaviourism and topic-neutral analysis, and the topic as we know it is decimated. Further remove all the reflections rooted in our manifest situation, and little or nothing would remain. After all, the idea of philosophy of mind as a distinct area within the wider discipline emerged only with physicalism (Rorty, 1982).

Why should this matter? Surely our only interest is in having an account of the nature of mind which is true. Two reasons why it matters are as follows. The first is that physicalism's consequences for consciousness are thoroughly problematic. Physicalism could be true if bare judgment can account for our manifest situation. But only physicalism suggests that it can, and physicalism is badly motivated; mainly by inattention to metaphilosophy. Moreover, if judgments about consciousness were indeed evidentially baseless, this in itself would cast serious doubt on physicalism.

The second reason is that this debate is occurring at a time when the science of consciousness and artificial intelligence research is racing along — hardly a coincidence. Before long, we may have machines that, from the outside, appear to have minds just like ours. If they pull sad faces or make sad noises when we go to switch them off, we will have qualms. Now all philosophers want to account for the manifest situation. But the way physicalist illusionism does this provides an ideal justification for allocations of machine consciousness. For we can program machines to represent the same kinds of physical things as us, and also to report phenomenal illusions; so if we believe that representation is all that is required to create the illusion of a manifest situation, our natural inclination to think they are like us so long as they act like us receives an intellectual stamp of approval. Without physicalism, we cannot know they share our manifest situation, because we cannot know that bare judgment is enough — *they* are making the judgments, so we cannot know how things seem to them, only how they tell us they seem. With physicalism, however, they must be conscious — because physicalism leads inexorably to illusionism.

This matters because, if these machines are invented, they will have a profound influence on human life. Maybe it will be good, maybe bad; but how we should treat them, and whether indeed we should press on in inventing them, is a debate we need to have. Physicalism removes philosophy's voice from that debate. It removes the rational basis from appeals to our manifest situation; the intuitive way we have of thinking about our lives and world. With physicalism as an unquestioned background, there is no room to question whether they will be genuinely like us when their behaviour and programming is objectively indistinguishable. With the conceptual space of transcendence closed down, this distinction cannot be made without disputing, rather than interpreting, the scientific worldview; and the latter is all philosophy can credibly do. So unless philosophy rapidly gets its act together, it will have no significant part to play in this evidently philosophical debate. Essentially, there will be no debate. Rather, anti-physicalist philosophers will be relegated to the sidelines, along with religious leaders, as the loony 'anti-science' faction, while the scientists developing the technology will do their best to make it safe... and then we will all just have to wait to see what happens when the machines join us. The idea that there might be a distinctively philosophical perspective to be had on these issues, besides cheering or smearing science, will be passed over. This is what is at stake in illusionism: philosophy's voice in a debate where it might be persuasive and important.

Illusionism is a sound and important metaphysical insight, which undermines the pseudo-challenge of phenomenal realism to science. But it can be taken in different directions, and my suggestion is to pursue it outside of the self-imposed shackles of physicalism. Frankish says that 'evolutionary theorizing about consciousness can flourish, once freed from the metaphysical preoccupations of [phenomenal] realism' (p. 29). I do not think serious scientific theorizing about consciousness should give two hoots about metaphysics; if it does, then this is the result of either scientists dipping into debates that are not their concern, philosophers trying to exert undue influence on science, or — what I think is actually the case — both. Philosophers should stay aware of how scientific debates go, so as to inform their work on the status of our objective representation of reality. But they should not try to take part in them; not in their day-jobs, at least. I am all for joined-up thinking across disciplines, but not if this results in philosophy being incapacitated and science being distracted by philosophical problems it cannot solve. So I prefer Frankish's sentence in an amended form: philosophical *and* scientific theorizing about consciousness can flourish, once freed from the metaphysical preoccupation of physicalism.

References

Chalmers, D.J. (1996) *The Conscious Mind*, New York: Oxford University Press.
Dennett, D.C. (1991) *Consciousness Explained*, Boston, MA: Little, Brown and Co.
Frankish, K. (this issue) Illusionism as a theory of consciousness, *Journal of Consciousness Studies*, **23** (11–12).
Jackson, F. (1982) Epiphenomenal qualia, *Philosophical Quarterly*, **32** (127), pp. 127–136.
Kripke, S. (1972) Naming and necessity, in Harman, G. & Davidson, D. (eds.) *Semantics of Natural Language*, pp. 253–355, Dordrecht: Reidel.
Loar, B. (1990) Phenomenal states, in Tomberlin, J. (ed.) *Philosophical Perspectives, 4: Action Theory and Philosophy of Mind*, pp. 81–108, Atascadero, CA: Ridgeview.
Nagel, T. (1974) What is it like to be a bat?, *Philosophical Review*, **83** (4), pp. 435–450.
Place, U.T. (1956) Is consciousness a brain process?, *British Journal of Psychology*, **47** (1), pp. 44–50.
Quine, W.V.O. (1975) A letter to Mr. Ostermann, in Bontempo, C. & Odell, S. (eds.) *The Owl of Minerva: Philosophers on Philosophy*, pp. 227–230, New York: McGraw-Hill.
Rorty, R. (1982) Contemporary philosophy of mind, *Synthese*, **53** (2), pp. 323–348.
Searle, J.R. (1997) *The Mystery of Consciousness*, London: Granta Publications.
Smart, J.J.C. (1959) Sensations and brain processes, *Philosophical Review*, **68** (2), pp. 141–156.
Tartaglia, J. (2013) Conceptualizing physical consciousness, *Philosophical Psychology*, **26** (6), pp. 817–838.
Tartaglia, J. (2016a) Rorty's philosophy of consciousness, in Malachowski, A. (ed.) *A Companion to Rorty*, forthcoming, Oxford: Wiley-Blackwell.
Tartaglia, J. (2016b) Rorty and the problem of consciousness, in Leach, S. & Tartaglia, J. (eds.) *Consciousness and the Great Philosophers*, London: Routledge.
Tartaglia, J. (2016c) *Philosophy in a Meaningless Life*, London: Bloomsbury.
Tartaglia, J. (2016d) Is philosophy all about the meaning of life?, *Metaphilosophy*, **47** (2), pp. 282–303.

Keith Frankish

Not Disillusioned

Reply to Commentators

Abstract: *This piece replies to commentators on my target article in this issue, 'Illusionism as a Theory of Consciousness', building on the arguments offered there. It groups commentators together by their attitude to illusionism, classifying them as advocates, explorers, sceptics, and opponents. It expands on the case for illusionism, refines the position, and responds to objections.*

> There is nothing more deceptive than an obvious fact. (Arthur Conan Doyle, 1892, p. 80)

I am grateful to the commentary authors for their contributions. The aim of this special issue is to give the reader a sense of the potential of illusionism as an approach to consciousness, and the commentators do an excellent job of this, both those who defend the approach and those who challenge it. Each commentary deserves a far more detailed reply than there is space for here, so I shall concentrate on the most salient issues for the overall evaluation of illusionism and focus on points of disagreement rather than agreement. (Thus, if I say relatively little about a piece, this should not be taken to mean that I dismiss it; quite the opposite.) To make this reply a smoother read, I shall group similar commentators together, classifying them as *advocates, explorers, sceptics, and opponents.*

1. Advocates

I begin with a group of commentators who offer further arguments in support of illusionism.

Correspondence:
Keith Frankish, The Open University, UK.   Email: k.frankish@gmail.com

Daniel Dennett provides a characteristically robust statement of the case for illusionism as the default theory of consciousness, arguing that we should thoroughly explore the mundane possibility of illusion before turning to exotic theoretical positions, especially when the latter offer few, if any, empirical predictions. I could not agree more, and if I have been less robust in stating the case for illusionism, it is only for tactical reasons. Of course, opponents will say that the methodological principle to which Dennett appeals is not applicable in this case, since illusionism denies the existence of the very thing to be explained. But this is begging the question, which is precisely whether phenomenality is real or illusory. The explanandum is the thing we call 'conscious experience', where it is an open question whether this involves phenomenality or the illusion of it (compare the inclusive sense of 'consciousness' defined in Section 1.6 of the target article). Illusionists agree that we have a potent intuition that phenomenality is real, but they hold that the rational policy (at least given our current, rudimentary understanding of the neuroscience of consciousness) is not to trust it and to pursue an illusionist research programme. If the programme proves fruitful, then the realist intuition may loosen its grip on us (or we may loosen our grip on it).

Of course, at present illusionists can do little to make it seem plausible that this will happen. They can at best offer vague sketches of how the illusion of phenomenality might be generated, which are easily dismissed. But if truth is our aim, then we should be prepared to put our realist intuition to the test. The widespread reluctance to do this suggests that there may be non-epistemic concerns lurking in the background. Perhaps people worry that ceasing to trust the intuition would erode our sense of self or our sympathy for the suffering of others, and feel that we should hold onto it regardless of its truth. Such worries are, I think, misconceived, but they deserve detailed articulation and assessment. (I shall make some brief remarks later, in responding to Katalin Balog.)

Dennett also urges caution in framing illusionist hypotheses and warns against supposing that there is a clear-cut range of questions and theoretical options that can be identified in advance of detailed empirical work (as some passages in my target article may have suggested). I think these points are wholly salutary.

In his commentary, **Jay Garfield** attacks phenomenal realism, arguing that phenomenal properties would be unknowable, that introspection affords no good evidence for their existence, and that belief in them arises from mistaking properties of external objects for properties of the sensory systems by which we perceive them. In making these arguments he draws on Sellars and Wittgenstein, but he goes on to show that similar ideas have long been present in Buddhist philosophy. In particular, he outlines Vasubandhu's view that it is a misconception to think of experience as having dual subjective and objective aspects — a misconception that yields a doubly distorted view of the causal processes involved.

I am, of course, sympathetic to Garfield's arguments. There are points at which phenomenal realists will want to object (arguing, for example, that zombies do not share the same phenomenal beliefs as us and that we have a special kind of epistemic access to our phenomenal properties), but I shall not discuss these objections here (some relevant points are made in Section 3 of the target article). Instead, I shall offer a couple of general observations.

First, a comment on the nature of Garfield's illusionism. If I read him right, Garfield sees phenomenal realism as a purely *cognitive* illusion, which consists in the mistaken belief that our experiences have phenomenal properties. This may be correct, but there might also be a quasi-perceptual element to the illusion. It is possible that we have sensory systems that target aspects of our brain activity, and that these systems play a role in generating the illusion of phenomenality (as proposed in Humphrey, 2011, for example). I take it that this possibility is compatible with Garfield's arguments, and indeed with rejection of subject/object duality. Such neuro-senses would be on a par with the other senses, including other body-directed ones such as proprioception. The properties they detect would be inner only in a spatial sense, would not be immediately and infallibly known (and would be known by zombies), and would not be phenomenal in any substantive sense (though they might be *represented* as phenomenal). I don't think we should rule out the possibility that such neuro-senses play a role in consciousness.

Second, a comment on Garfield's discussion of Buddhist philosophy. Garfield has done Western philosophers a tremendous service in introducing them to Buddhist philosophical traditions, which, as he shows, contain much of great contemporary interest and importance (see in particular Garfield, 2015). I am not qualified to comment in detail on the points he makes, but the fact that illusionist ideas can be found in ancient Buddhist philosophy is in itself significant. Illusionists are sometimes accused of *scientism* — as if only blind science worship could prompt someone to deny the existence of phenomenal properties. I think this is unfair, and the fact that similar views emerged in a quite different intellectual culture long before the development of modern science helps to rebut it. Vasubandhu's illusionism was the product of a long tradition of metaphysical reflection on the nature of the world and our place in it, and the fact that many Western philosophers find illusionism utterly implausible may say more about their cultural horizons than about the nature of consciousness itself.

Georges Rey devotes his commentary to exploring the nature and origin of the intuition that underlies phenomenal realism. He distinguishes w(eak)-consciousness, which involves implementing various computational processes of attention and internal awareness, and s(trong)-consciousness, which involves meeting some additional, non-computational condition (these notions correspond to the two senses of what-it's-like-ness distinguished in Section 1.7 of the target article). We have a powerful intuition that we have s-consciousness, but we have no idea what the extra condition might be nor any independent test for its presence. It is often assumed that we have rationally compelling introspective grounds for believing in s-consciousness, but Rey questions this. Given the elusiveness of the condition and the known fallibility of introspection, there is scope to doubt that we really have s-conscious states, as opposed to merely having the attitudes and reactions we associate with them. Moreover, Rey notes that we have an equally strong conviction that other people possess s-consciousness, suggesting that our concept of it is sensitive to behavioural factors (perhaps including marks of biological life) as well as to introspective ones. All these points are, I think, very well taken.

Rey goes on to offer a Wittgensteinian diagnosis, according to which talk of 'consciousness' (in the strong sense), like that of 'the sky', has a role within a particular everyday linguistic practice, or 'language game', which cannot be smoothly integrated with science. The concept plays a useful role, reflecting everyday needs, interests, and moral concerns, but we cannot specify its conditions of application, and it does not appear to pick out a well-defined natural phenomenon. (I would add that the fact that we cannot specify its conditions of application means that we cannot be sure that they are *not* wholly functional and behavioural.)

I think Rey's diagnosis is useful and that understanding the nature and function of the concept of s-consciousness will be crucial to developing the illusionist case. Our intuitions here may be put to the test in the not too distant future, as we create humanoid robots that have w-consciousness and exhibit a rich variety of human-like behaviour (enabling us to interact with and control them using our existing social skills and knowledge). I suspect such machines will provoke conflicting intuitions. When we interact with them, they will strongly activate our concept of s-consciousness, but when we reflect on how they were made and how they work, we shall have a strong intuition that they lack s-consciousness. This may lead to widespread scrutiny of the concept itself, and perhaps to its revision or replacement.

Rey concludes with some cautionary remarks: some aspects of experience (especially of colour experience) seem deeply resistant to illusionist explanation, and the intuition that s-consciousness is real remains tenacious. It is important that illusionists say these things. They do not claim to have explanations for specific features of conscious experience, or even to see how such explanations will go. They simply claim that the illusionist programme is the most promising one, and that our current intuitions about what can and cannot be explained in illusionist terms may not be reliable. Detailed empirical work may open new theoretical and conceptual options. We may never fully dispel the illusion of s-consciousness; it may be hardwired into our mechanisms of introspection and social perception, just as some visual illusions are hardwired into our visual systems. But recognizing that that is the case will be a major step forward.

One final point: Rey notes that there is a passage in the target article where I speak of phenomenal properties, conceived as non-existent intentional objects, as being causally potent (Rey, this issue, p. 201, fn. 9). Rey dissociates himself from this view: it is the representations of non-existent intentional objects that are causally efficacious, and talk of the objects themselves having certain effects is merely a convenient shorthand. In fact, I agree with Rey on this; the passage in question was loosely phrased.

Amber Ross highlights some epistemological problems for phenomenal realism. Real properties are independent of our beliefs about them (reality, in Philip K. Dick's words, 'is that which, when you stop believing in it, doesn't go away' (Dick, 1995, p. 261) — and, we might add, real properties are ones that don't come to be there just because you think they are). If phenomenal properties do not exhibit this sort of belief-independence, then the natural conclusion is that they are not real but merely intentional objects of our representations, like the content of a fiction or hallucination. As Ross puts it:

> Any view according to which the subject's beliefs about the character of her conscious experience do play a role in determining the facts of the matter about her conscious experience is a non-realist, illusionist type of view. (Ross, this issue, p. 221)

Yet, as Ross shows in some detail, it is very hard to describe a plausible scenario in which a subject has a false belief about the phenomenal character of an experience they are currently attending to. Of course, it wouldn't exactly help the realist if we could construct such a scenario; for, as Ross notes, we have a strong intuition that we cannot make this kind of mistake (this issue, p. 219). In this respect, then, illusionism is *better* placed to account for the common-sense view of consciousness than phenomenal realism.

I think the line of attack Ross pursues — questioning the coherence of phenomenal realism — is an important one for the illusionist, and one that was perhaps insufficiently stressed in the target article. It is, of course, a line that Dennett has pressed with considerable force over the years. One example he uses is that of change blindness (Dennett, 2005, chapter 4). People can fail to notice repeated shifts of colour in an image, provided each presentation of the image is separated by a brief masking stimulus. In such cases, the colour shifts must be registered at some level by the subject's visual system, but do they show up in their visual phenomenology? Dennett argues that realists face a dilemma: if they were to experience change blindness themselves, what would they say about their own phenomenology?[1] If they would say that their phenomenal properties changed without their noticing it, then they must accept that we are not authoritative about our phenomenal properties and that, for all we know, they may change all the time without our noticing. If they say that their phenomenal properties did not shift until they noticed the change in the image (that is, registered it cognitively), then it looks as if our phenomenal properties are simply constructions out of our judgments, as illusionists claim (and if they say they don't know if their phenomenal properties shifted, then it is unclear what could possibly settle the matter). The upshot, Dennett concludes, is that the notion of a phenomenal property is simply a mess, a source of nothing but confusion. Ross's contribution illustrates the force of considerations like this, which are, I think, still widely underestimated.

---

[1] Dennett uses the term 'qualia', but for consistency I'll put the point in terms of phenomenal properties.

James Tartaglia advocates a surprising position: non-physicalist illusionism. I did not consider the possibility of such a position in the target article, since I was concerned with illusionism as a conservative explanatory strategy, but of course illusionism does not *entail* physicalism. One could be an illusionist about consciousness while holding that reality is fundamentally non-physical. Indeed, Tartaglia argues that illusionism actually provides grounds for holding that.

In the first part of his commentary, Tartaglia attacks 'intermediate' positions, which attempt to combine physicalism with phenomenal realism. Such positions typically rely on the phenomenal concept strategy, but, Tartaglia argues, this reliance is unwise, since phenomenal concepts aren't simply neutral ones, which do not present their objects as physical, but substantive ones, which present their objects as having a qualitative, subjective nature that no physical property could have. If physicalism is true, then phenomenal concepts must misrepresent their objects:

> So if the phenomenal concept is a concept of a brain state, it must be a radical misconception of it; we must be misconceiving the brain state beyond all recognition, in fact. We are thinking of a brain state as a subjective experiential array, but that is not what it is at all. Consequently, the array must be an illusion, even if thinking about it somehow allows us to think about real brain states. (Tartaglia, this issue, p. 238)

Tartaglia has no sympathy with dualist or panpsychist explanations of consciousness, and he accordingly adopts an illusionist position. This line of argument is, of course, one that I endorse, and Tartaglia's presentation of it is elegant and compelling.

In the rest of his commentary Tartaglia turns to the metaphysical implications of illusionism. He points out that our metaphysics should be compatible with our manifest situation — with how things seem to us, and specifically with the fact that we seem to be confronted with arrays of phenomenal properties. Physicalist illusionists explain our manifest situation by appealing to our judgments and representations: things seem that way because that is how we are inclined to judge them to be. Tartaglia thinks this has profound epistemic consequences. In particular, he argues that it means we cannot be confident in our physical conception of reality, since that conception was created on the basis of illusory experience. He is not suggesting that we should doubt our science. By taking experience as a guide to reality, we have built up a coherent and detailed picture of an objective world — a picture that has eventually led us to the hypothesis that experiences themselves are illusory. But, he argues, we should not forget our starting point: experience itself has a reality that must be accounted for. Hence, we must supplement our scientific picture of reality with a distinctively philosophical account. There must be an independent reality behind our experience, which transcends the objective world in the same way that the objective world transcends the world of a dream (however, Tartaglia denies that this independent reality is mental; his view is not a form of idealism or phenomenal realism — this issue, p. 251, fn. 10).

What should we make of this? Does illusionism require us to reject physicalism? I am unpersuaded. I place myself in the tradition of Quinean naturalism, which (as Tartaglia notes) denies a sharp distinction between philosophy and science and holds that science, broadly construed, provides our best picture of reality. At any rate, I do not see how positing a transcendent reality could shed any further light on the nature of consciousness and subjectivity. (Tartaglia himself says that we can say 'nothing substantive' about independent reality; this issue, p. 250.) Tartaglia worries that physicalist illusionism renders many other aspects of our manifest situation illusory too, including our sense of being spatio-temporally located. Even if this were so (and I'm not sure it is), I do not see it as a reason to reject physicalism. If certain illusions are important to us, we can continue to live by them, treating them as enabling fictions, which do not need metaphysical underpinning.

Moreover, I think Tartaglia overestimates the negative epistemic implications of illusionism. He suggests that it renders our judgments about our experiences 'completely unreliable' (this issue, p. 244). But this is too swift. Illusionism does not claim that our conscious experiences are wholly illusory, only that their apparent *phenomenal aspect* is. I may be correct to judge that I am currently seeing a red postbox (where red is a reflectance property of surfaces), even though I'd be wrong to judge that the experience has a reddish phenomenal feel. Tartaglia asks how we can correlate experiences with worldly properties, but I fail to see the problem. Evolution has set up the correlations, designing our perceptual systems to reliably track worldly properties (perhaps disjunctive, gerrymandered ones), and by doing science we can get a better understanding of the nature of those properties. Tartaglia doubts that we can 'cherry-pick' experience for veridical elements, but that is just what the scientific method has enabled us to do. Even if it is not the fundamental reality, the objective world has a complex structure independent of us, and by applying the scientific method we have acquired a powerful grip on that structure. In the process, we have come to question some of the beliefs we started with, but unless we endorse some form of foundationalism, this is not a problem.

In the same vein, illusionists need not deny that phenomenal concepts represent real properties, albeit under distorted guises. Tartaglia finds it implausible that phenomenal concepts represent properties either of brain states or of distal objects, arguing that the misrepresentation involved would be too radical (this issue, pp. 244–5). Again, I fail to see the worry. A concept may track a certain property even if it radically misrepresents it. Recall Humphrey's example of the Penrose triangle, discussed in the target article (Frankish, this issue, p. 17). Deployed in visual experience, the concept of a Penrose triangle tracks a certain sort of three-dimensional structure (a 'Gregundrum'), which it represents as a physically impossible object. The misrepresentation involved is radical yet perfectly possible. Moreover, it may be useful if we need to distinguish Gregundra from non-Gregundra. Gregundra are simply objects that create the illusion of a Penrose triangle, and the best way to tell if an object is a Gregundrum is to see if it creates the illusion — if our visual system misrepresents it as a Penrose triangle. It would be impossible to develop a veridical perceptual concept that reliably distinguishes Gregundra from all the other similar structures that do not create the illusion. Something similar may be the case with phenomenal concepts. They may pick out highly disjunctive physical properties (either of our cognitive systems or of distal objects) which it is useful for us to track but which are unified only by the fact that they trigger the concept. The fact that they radically misrepresent their objects is no bar to their performing this function.

Perhaps my Quinean sympathies — or my lack of what Tartaglia calls 'historically-informed metaphilosophical self-consciousness' (this issue, p. 252) — are blinding me to Tartaglia's deeper point. At any rate, as far as the science of consciousness goes, he and I are in agreement: we should adopt an illusionist view.

2. Explorers

I turn now to four commentators I have dubbed *explorers*. These use their commentaries to explore ways of developing illusionism — either building theories, responding to objections, or reviewing experimental evidence. Their papers illustrate how illusionism can form the core of a research programme, which can be supplemented and developed in different ways.

François Kammerer addresses the illusion problem — the problem of explaining how the illusion of phenomenality arises. As he notes, the problem has a particularly hard aspect. It is not just that we are strongly disposed to think that phenomenality is not an illusion; we find it hard to understand how it *could* be an illusion:

> [W]hat makes us reluctant to accept illusionism is not only that we are disposed to believe that we are conscious, it is also that we have difficulties *making sense of the hypothesis that we are not conscious while it seems to us that we are.* (Kammerer, this issue, p. 127)

This, Kammerer argues, sets this illusion of phenomenality apart from all other illusions and means that it cannot be usefully modelled on them.

Kammerer proposes a solution. Simplified somewhat, it runs as follows.[2] Introspection is informed by an innate and modular theory of mind and epistemology, which states that (a) we acquire perceptual information via mental states — experiences — whose properties determine how the world appears to us, and (b) experiences can be fallacious, a fallacious experience of A being one in which we are mentally affected in the same way as when we have a veridical experience of A, except that A is not present. Given this theory, Kammerer notes, it is incoherent to suppose that we could have a fallacious experience of an experience, E. For that would involve being mentally affected in the same way as when we have a veridical experience of E, without E being present. But when we are having a veridical experience of E, we are having E (otherwise the experience wouldn't be veridical). So, if we are mentally affected in the same way as when we are having a veridical experience of E, then we are having E. So E is both present and not present, which is contradictory. (Kammerer couches the argument in terms of experiences, but it could easily be recast in terms of the phenomenal properties of experience. Having a fallacious experience of a phenomenal property involves being mentally affected in the way one would be if the property were present, which involves it being present. Generalized further, this argument might explain our sense that introspection is infallible.)

---

[2] I have omitted a lot of Kammerer's detail, but I hope I have captured the core of his argument.

Kammerer proposes that this explains the peculiar hardness of the illusion problem. The illusionist thesis cannot be coherently articulated using our everyday concept of illusion, which is rooted in our naïve concept of fallacious experience. Moreover, if the naïve theory Kammerer sketches does inform our introspective activity, then we shall not be able to form any imaginative conception of what it would be like for illusionism to be true. Hence the common claim that, where consciousness is concerned, appearance is reality. As Kammerer stresses, this does not mean that illusionism actually is incoherent. It simply means that in order to state it we must employ a technical concept of illusion — as, say, a cognitively impenetrable, non-veridical mental representation that is systematically generated in certain circumstances.

Kammerer's approach to the illusion problem is, I think, a promising one, and the idea that introspection is theoretically informed is likely to figure prominently in any developed illusionist theory. Of course, even if Kammerer is right about the source of our intuitive resistance to illusionism, this would not show that illusionism is true, though it would help to dispel one common objection to it. Realists will say that phenomenality is not an illusion even in a technical sense: our relation to our phenomenal properties is one of direct acquaintance, which does not depend on potentially fallible representational processes. Perhaps Kammerer could employ the strategy again here, arguing that our concept of introspective acquaintance is also a theoretical one. At any rate, considerations like this should help to move the debate forward, beyond the simple assertion that illusionism is unintelligible.

Derk Pereboom has done much to establish illusionism as a respectable approach to consciousness, setting out a carefully articulated illusionist theory (the 'qualitative inaccuracy hypothesis') and showing how it can be used to rebut standard anti-physicalist arguments (see Pereboom, 2011). In his commentary, he discusses the form an illusionist theory should take, challenging the functionalist view I suggested in the target article. He makes two points. The first concerns what illusionists should say about illusions of phenomenality themselves. Realists will object that illusionists are still committed to phenomenal realism, since there is something it is like to have the illusion of a phenomenal property. In the target article, I suggested that illusionists should deny that phenomenal illusions themselves seem to have phenomenal properties. Pereboom thinks this won't do, and argues that we should instead explain their apparent phenomenality as a further illusion. If introspection misrepresents quasi-phenomenal states as phenomenal, then it can misrepresent our modes of presentation of those states as phenomenal too. This need not create a regress, Pereboom argues, since there is no reason to think that we also represent those higher-order modes of presentation, or at least that we do so under phenomenal modes of presentation.

It is good to have this proposal on the table. I think it is a coherent position, and I agree with Pereboom that the regress objection is not serious (the point to stress is that mental states seem to possess phenomenal properties only when introspected, and psychological limitations on the introspection process will naturally block the regress). It is in some ways a puzzling proposal, however. Why should introspection represent modes of presentation as having the same properties as the states they represent? Why should the representation of an experience feel like the experience itself? Moreover, we may not need to posit higher-order introspective processes in order to account for our sense that illusions of phenomenality would themselves have phenomenal properties. Recall Kammerer's proposal about the theory-laden nature of introspection. If Kammerer is right, then when we try to conceive of an introspective illusion we shall conceive of a mental state that incorporates the original experience, with all its (apparent) phenomenal properties. The apparent higher-order feel may simply be an artefact of the innate theory that informs introspection.

Pereboom's second point concerns the nature of quasi-phenomenal properties. Introspection represents these as intrinsic properties rather than functional ones. Illusionism removes the pressure to think of them in this way, allowing us to incorporate them smoothly into a functionalist account of the mind. But, Pereboom argues, illusionism doesn't *require* us to adopt a functionalist view. We could regard quasi-phenomenal properties as consisting, at least partially, of *absolutely intrinsic aptnesses*, which form the categorical bases for the causal powers of physical entities. Pereboom argues that this view not only vindicates our common-sense intuition that phenomenal properties are intrinsic causal powers but also gives the illusionist a stronger response to standard anti-physicalist arguments.

Again, it is good to have this view on the table, though personally I find it unpersuasive. Of course, if one thinks that all causal powers are ultimately grounded in absolutely intrinsic aptnesses, then one will think that the powers of quasi-phenomenal properties are too. But I don't think there are *specific* reasons for thinking of consciousness in this way. As Pereboom acknowledges, we do not need to posit absolutely intrinsic properties in order to rebut the anti-physicalist arguments, and our sense that phenomenal properties are intrinsic ones can be explained as a misrepresentation. As illusionists, we do not need the heavy metaphysical machinery of absolutely intrinsic aptnesses in order to explain why conscious experiences seem to be intrinsic causal powers, and employing it would, to my mind, bring illusionism uncomfortably close to a form of Russellian monism.

I turn now to two contributions from scientists. Much scientific work on consciousness has been conducted, wittingly or not, in a quasi-dualistic spirit. Theorists seek to identify the neural processes that produce consciousness, without offering any explanation of *how* they produce it. This isn't surprising if consciousness is conceived in a realist way: it is impossible to gain any explanatory purchase on such a nebulous phenomenon. But illusionism provides a much more tractable target for scientific investigation. To explain consciousness we need to identify and explain the (broadly representational) processes that collectively constitute the illusion of phenomenality. The two commentaries considered next adopt this perspective.

Michael Graziano provides a clear introduction to his *attention schema theory*, according to which consciousness depends on possession of an internal model of one's attentional processes. Graziano's conception of the explanandum for a theory of consciousness is thoroughly illusionist. As he explains:

> Here by 'consciousness' I mean that, in addition to processing information, people report that they have a conscious, subjective experience of at least some of that information. The attention schema theory is a specific explanation for how we make that claim... It is a theory of how the human machine claims to have consciousness and assigns a high degree of certainty to that conclusion. (Graziano, this issue, p. 98)

The aim is to explain our sense that we are conscious, rather than consciousness itself as a distinct property. This sense arises, Graziano argues, from the fact that, in addition to representing features of the world and of ourselves, we represent our *mental relation* to things via attention. We have an 'attention schema', which models covert attention (the deep processing of selected information), allowing us to monitor and control it. This model does not provide a detailed representation of the mechanisms involved; rather, it represents attention in an abstract, schematic way, as a sort of private *mental possession* of something. As a result, when we introspect our attentional processes we seem to find an inner world where a subjective self has an immediate grasp of the properties of things, leading us to issue reports like this (which Graziano puts into the mouth of a robot equipped with an attention schema):

> 'my mental possession of the apple, the mental possession in-and-of-itself, has no physically describable properties. It's an essence located inside me... It's my mind taking hold of things — the colour, the shape, the location. My subjective self seizes those things.' (Graziano, this issue, pp. 102–3)

When we talk of consciousness and its features, we are reporting the deliverances of our attention schema. Graziano goes on to outline experimental evidence that consciousness is associated with the control functions of the attention schema, as the theory would predict.

I shall not attempt to assess attention schema theory here but simply comment on its relation to the illusionist programme. Graziano notes the affinity (especially with Dennett's views) but argues that the term 'illusionism' has misleading connections. To describe consciousness as an illusion suggests that it is nothing at all and that introspection is simply in error. But, he points out, covert attention itself is real, and our internal model of it, though schematic and abstract, is well adapted for its function of tracking and controlling attention. As he puts it, 'consciousness is not an illusion but a useful caricature of something real and mechanistic' (this issue, p. 112).

I think these are excellent points, and they indicate the need for an important clarification. Talk of illusion does double duty within illusionist theorizing. On the one hand, it may refer to *quasi-perceptual* introspective representations generated by self-monitoring processes, such as the attention schema. These representations may be highly abstract and distorted, and in that sense illusory, but they may also carry valuable information for the system and facilitate important tasks of control and self-manipulation. An illusion need not be a fault and may have been carefully designed (compare Dennett's analogy with the 'user illusions' produced by the icons and pointers on a computer desktop — Dennett, 1991). On the other hand, illusion talk may refer to the *cognitive* illusion involved in judging that we are acquainted with an internal world of intrinsic phenomenal properties. Here it is appropriate to talk of error (certainly in theoretical contexts), though perhaps still not of a fault: belief in the metaphysical specialness of our inner lives may be adaptive, playing an important role in human psychology and social interaction (Humphrey, 2011).

These illusions, quasi-perceptual and cognitive, are of course closely related; we judge that we are acquainted with phenomenal properties because introspection gives us such a partial view of internal reality (indeed, natural selection may have sculpted our neural processes in order to create the cognitive illusion; *ibid.*). Phenomenal consciousness, we might say, is a theoretical illusion built on an introspective caricature.

In their commentary, **Nicole Marinsek** and **Michael Gazzaniga** look at illusionism from the perspective of split-brain research. Patients who have undergone surgical severing of the corpus callosum display various behavioural dissociations, which suggest that each hemisphere is operating as a separate mind. This presents a challenge for illusionism. Both hemispheres show signs of being phenomenally conscious (in the everyday sense), so if phenomenality is an introspective illusion, then both must possess a capacity for introspection and be susceptible to illusions. Marinsek and Gazzaniga review relevant experimental evidence and tentatively conclude that this is indeed the case. One moral of this, they suggest, is that, even without callosotomy, phenomenal consciousness may be fragmented, comprising numerous 'modular illusions' with different characteristics.

I think these points are well taken, and, as Marinsek and Gazzaniga note, the split-brain literature will provide a useful testing ground for detailed illusionist proposals (it would be interesting to explore its implications for attention schema theory and for Humphrey's 'sentition' theory). Moreover, the suggestion that consciousness may be fragmented is, I think, an important one. One thing the split-brain literature has shown is that our sense of psychological unity can be illusory: despite the dissociations in their behaviour, split-brain patients continue to feel unified, and they unconsciously confabulate to preserve that feeling. Of course, if we conceive of subjecthood in non-psychological terms, as involving direct acquaintance with phenomenal properties, then it is hard to see how we can establish any objective criteria for identifying conscious subjects. But illusionism provides a much more tractable approach. To be a conscious subject is (putting it very sketchily) to be a system that produces appropriate introspective representations of its own mental activity and uses them to modulate its activity in appropriate ways. In this sense, we may each incorporate multiple conscious or semi-conscious subjects, either modular or partially integrated with each other.

3. Sceptics

This section looks at contributions from four commentators who, although not full-blown opponents of illusionism, express reservations about the position or feel that it is in some way misguided.

Susan Blackmore distinguishes illusionism from a more cautious view, which she calls *delusionism*. Whereas illusionists deny the existence of phenomenal consciousness outright, delusionists hold that we have many mistaken theories about it. Blackmore expresses reservations about illusionism, but she endorses delusionism, arguing that we are wrong to think that there is a stream of consciousness, with rich, unified, and determinate contents, and a persisting self, which observes it. Whenever we introspect, we always find some conscious experience, and this leads us to think that there is a continuous inner stream of such experiences and an inner self waiting to observe them. But these claims, she argues, are baseless — neither neuroscience nor careful introspection offers any way of determining whether conscious experiences are present at times when we are not actively introspecting. Rather, there are just moments of consciousness, temporary constructions bonding thoughts and perceptions to a representation of the self.

This is a valuable piece, which usefully summarizes ideas that Blackmore has defended at length in earlier work. There are many important issues here, but I shall confine myself to commenting on the relation between delusionism and illusionism. Illusionism clearly entails delusionism: if there are no phenomenally conscious experiences, then there is no continuous stream of them either. Could there be a continuous *illusion* of consciousness? It depends on what kind of illusion we are thinking of. If it is a personal-level cognitive one, which occurs when we actively introspect and judge that we are currently having an experience with such-and-such phenomenal properties, then the answer is obviously no. But illusionists might want to say that there is a continuous subpersonal illusion, or something like it, consisting in the production of abstract, quasi-perceptual representations of neural processes, which are used for internal control purposes and which form the basis for our phenomenal judgments when they occur (perhaps Graziano's attention schema theory supposes something like this).

What about the converse? Does delusionism entail illusionism? Blackmore thinks it does not. She does not endorse illusionism and seems to accept the reality of immediate conscious sensations. *Prima facie* this seems right — there could be moments of phenomenal consciousness without a continuous stream of it. However, I am not sure this positon is stable. If there are such moments, then there are properties of one's brain state at those moments that make it phenomenally conscious — physical properties, let us assume. But then it should be possible, at least in principle, to determine whether our brain states have these properties at times when we are not introspecting, and thus to determine whether or not there is a stream of consciousness. If delusionists deny that this is possible, then, it seems, they should deny that there are such properties and accept that phenomenal consciousness does not exist.[3]

Blackmore closes her commentary by suggesting that our delusions of consciousness are malign memes, which we can, with effort, rid ourselves of. I am unsure about this. It may be true that our conceptions of consciousness and the self are culturally shaped, though rooted in the deliverances of real introspective processes. However, they may not be malign — they may play valuable social and psychological roles, as Humphrey has argued (Humphrey, 2011). As we understand more about why we conceptualize our inner lives in the way we do, we should gain more purchase on these questions, perhaps with beneficial practical consequences.

Nicholas Humphrey uses his commentary to question the value of characterizing consciousness in terms of illusion. In the past, Humphrey has proposed an explicitly illusionist theory, according to which conscious experiences reflect internalized expressive responses to stimuli, which interact with incoming sensory signals to generate complex feedback loops. When these loops are internally monitored, Humphrey argued, they appear to possess strange qualitative and temporal properties, creating the illusion of a magical inner world (*ibid.*).

In his commentary, however, Humphrey repudiates the label 'illusionist' and insists that his view is better characterized as a realist or 'surrealist' one (though not, he stresses, in any anti-physicalist sense). He offers two reasons for this. One is tactical: to characterize one's view as the claim that phenomenal consciousness is an illusion

---

[3] There are passages in Blackmore's commentary which suggest that her sympathies are more illusionist than she admits. She writes, for example, that neuroscientists 'will never find the neural correlates of an extra added ingredient — "consciousness itself" — for there is no such thing' (this issue, p. 61).

is to invite people to ignore or ridicule it; it's 'bad politics' (this issue, p. 122). I shall discuss this worry in a moment. Humphrey's other, more substantive, point is that sensations represent something real and important — namely our evaluative responses to stimuli:

> [W]hen considering whether sensations are or are not 'real', we must never let go of the fact that sensations do indeed represent *our take* on stimuli impinging on the body. In doing so they represent some of the objective facts about what's happening: the what, where, and when, for example. But, crucially, they also represent how we *evaluate* what's happening, how we *feel* about it. And this is where phenomenal properties come into their own. Sensations represent how we relate to stimulation using, as it were, a paintbox of phenomenal concepts to depict what it's like for us. (Humphrey, this issue, p. 118)

Sensations, he argues, represent two aspects of stimulation: how we are being stimulated (the objective side) and how we respond to the stimulation (the subjective side). Their phenomenal aspect corresponds to the latter — it represents our subjective take on stimulation. And this aspect, Humphrey argues, cannot be illusory or non-veridical:

> How could you... be experiencing a feel that 'doesn't exist'? To be blunt, I think the very notion of this is absurd. When the sensation represents you as feeling a certain way about the stimulation, *that is all there is to it.* The phenomenal feel arises with the representation, and *thereby its existence becomes a fact.* (Humphrey, this issue, p. 119)

There are two ways of reading this. On one, Humphrey is making a point similar to Graziano's: sensations are not mere illusions but representations of something real and important — our evaluative responses to stimuli, what they mean for us. (It is interesting that both Graziano and Humphrey hold that consciousness is based in a dynamic relation rather than passive awareness. Dennett makes a similar point in his commentary.) This reading is compatible with illusionism in my sense. For *phenomenal properties* may still be illusory. It may be that sensation misrepresents our evaluative responses (which are constituted by complex patterns of efferent neural activity) as simple intrinsic phenomenal feels — that it tells us how we feel in the language of phenomenal fictions. Again, this need not imply any *fault* in sensation. The distortion may be necessary to achieve the effect; the representation of a huge swathe of neural activity wouldn't have the same impact as a representation of phenomenal pain, just as a pile of sociological reports on parent–child relations wouldn't have the same impact as a performance of *King*

Lear. The analogy is with a skilled magician producing astonishing effects, not a desert traveller hallucinating an oasis. (I would add that, *pace* Humphrey, *genuine* error may be possible on the subjective side as well as the objective one. Introspection may sometimes go awry, representing the presence of an internal response that has not in fact been triggered.)

On the other reading, Humphrey is claiming that representations of our evaluative responses to stimuli *create* or *constitute* phenomenal feels as distinct properties. This, of course, is not an illusionist position but a realist one. There are places in the commentary where Humphrey appears to endorse this view, describing phenomenal properties as an 'inherent feature' of brain activity (this issue, p. 120).

On the whole, however, I think the illusionist reading is more accurate. When Humphrey defends his realism, it is the reality of our *relation* to stimuli that he stresses, not the reality of phenomenal feels themselves — which are, after all, usually characterized as non-relational properties (this issue, pp. 119–20). And when Humphrey talks of phenomenal feels 'aris[ing] with their representation' he may mean that they are intentional objects, which are real *for the subject* — part of their represented inner world. I suspect, then, that Humphrey is still an illusionist at heart: introspection uses a 'paintbox of phenomenal concepts' to represent certain internalized responses that are not really phenomenal. This chimes well with his adoption of the term 'surrealism'. Surrealist paintings distort reality, albeit in a creative, expressive way.

Back to the bad politics worry. Should illusionists adopt a less provocative label for their position? 'Illusionism' does have some undesirable connotations (any single-word label would have — 'magicism' would be even worse!). And, as I have stressed, we can employ an inclusive concept of consciousness that does not carry a commitment to phenomenal realism and allows us to affirm the reality and significance of consciousness in a natural way. But as a term for a theoretical approach to consciousness, I prefer to stick with 'illusionism' — at least at this stage in the debate, where the ghost of phenomenality has yet to be exorcised from cognitive science. It is all too easy to adopt a conception of what needs to be explained that encourages scientists and philosophers to ask bad questions and to ignore good ones. In my view, we need to challenge this misconception head on. There's no point mincing words: we don't have phenomenal properties, only representations of them.

Pete Mandik expresses a sympathetic scepticism about illusionism, agreeing that we do not have phenomenal properties but denying that we are under the illusion of having them. He makes two points. First, the term 'phenomenal' (as used in this context) has no clear content. Attempts to define it cycle uninformatively through a series of synonyms, and illusionists won't want to rely on private introspective ostension to explicate the concept. Mandik's second worry concerns the notion of illusion. Illusions are systematic: it is appropriate to talk of illusion only when a certain stimulus or scenario reliably evokes a certain misperception or fallacious judgment in people. In the case of consciousness, the illusion is supposed to be that when we introspect our experiences they seem to possess anomalous, inexplicable properties. But, Mandik notes, introspection elicits this judgment in few people outside philosophy departments — most would say their experiences seem perfectly mundane and natural. Mandik is tempted to call his position *meta-illusionism* but settles on *qualia quietism*: questions about qualia, or phenomenal properties, are simply not well enough defined to be worth pursuing.

I am sympathetic to Mandik's quietism (which echoes points Daniel Dennett has made), and to some extent I think the difference between us is one of emphasis. However, I think Mandik overstates his case. Take 'phenomenal' first. The concept is typically introduced via a sort of language game (call it 'the phenomenality language game'), which involves a combination of inner ostension (think of how pain feels, coffee smells, etc.),[4] reflection on the appearance/reality distinction (where is the colour of an after-image located?), thought experiments (imagine inverts and zombies), and scientific knowledge (science tells us that colours are really 'in' us), supplemented with theoretical claims (phenomenal properties are ineffable, intrinsic, radically private, and so on). This game isn't played only by philosophers; many of the moves are widely known, and children spontaneously invent some of them for themselves. I don't claim that the resulting concept is fully coherent, but it is not contentless either, and I think it is meaningful (and important) to deny that it picks out something real.[5] Moreover, I think there is a genuine introspective basis to the

---

[4] Illusionists can engage in inner ostension, though they will take the objects identified to be merely intentional.

[5] I do, however, doubt that there is any content to the weaker, 'diet' conception of qualia sometimes proposed, which is supposedly independent of the phenomenality language game — see Frankish (2012).

concept. It is sometimes argued that experience is wholly transparent — that we are aware only of aspects of the external world (including our bodies). But it is plausible to think that experience tells us more than this. As several commentators have argued, conscious experience also tells us about our *relation* to worldly properties — how we feel about them (Humphrey, this issue), or how we are attending to them (Graziano, this issue), or what expectations and reactions they evoke in us (Dennett, 2013). Introspection represents these relations under highly abstract, caricatured guises (creating what Dennett calls the *user illusion*), but in doing so it provides a substantive, though misleading, content to the notion of phenomenality.

Second, what about illusion? As I've mentioned, illusion talk does double duty — referring both to quasi-perceptual introspective representations of the sort just mentioned and to the cognitive illusion involved in judging that introspection acquaints us with phenomenal properties (and that these properties are anomalous or magical). The former might be better described as a caricature rather than an illusion, but I think the latter deserves the title. It is true, as Mandik observes, that mere introspection does not elicit these judgments; one needs to have been inducted into the relevant language game. But with that induction (which is common), and a little reflection, people do reliably endorse phenomenal realism and judge that it presents a hard problem. The case is similar to that of the Monty Hall problem, which Mandik cites as an example of a cognitive illusion. In order to fall for the illusion, one needs some prior (albeit imperfect) grasp of the concept of probability.

Finally, a point about quietism. From a tactical point of view, I think, quietism is not the best approach for the qualia irrealist. People easily fall into dualist ways of thinking, which lead them to ask bad questions about consciousness. To counter this, the irrealist needs to provide a robust explanation of why we are susceptible to dualist intuitions and why they seem so compelling. Simply telling them not to talk about qualia won't do. There is a Bob Newhart sketch in which he plays a psychiatrist whose only advice to a neurotic patient is 'Stop it!', repeated over and over. Quietism is a bit like that. It may be sound advice, but it's not very helpful therapeutically.

Eric Schwitzgebel also considers the problem of defining phenomenal consciousness, responding to the target article's challenge to identify a notion of phenomenal consciousness that is substantive yet free of dubious theoretical commitments. He proceeds by offering a definition by example, describing a range of uncontentious positive and negative cases and identifying phenomenal consciousness as 'the most folk-psychologically obvious thing or feature that the positive examples possess and that the negative examples lack' (this issue, p. 229). Because it relies on examples and does not adjudicate on contentious cases, this definition is a theoretically innocent one, yet it is not deflationary and leaves room for puzzlement about the nature of consciousness.

I think Schwitzgebel succeeds in identifying an important folk-psychological kind — indeed the very one that should be our focus in theorizing about consciousness. However, I don't think he has met the challenge of the target article. For, precisely because his definition is so innocent, it is not incompatible with illusionism. As I stressed in the target article, illusionists do not deny the existence of the mental states we *describe* as phenomenally conscious, nor do they deny that we can introspectively recognize these states when they occur in us. Moreover, they can accept that these states share some unifying feature. But they add that this feature is not possession of phenomenal properties (qualia, what-it's-like-ness, etc.) in the substantive sense created by the phenomenality language game. Rather, it is possession of introspectable properties that dispose us to judge that the states possess phenomenal properties in that substantive sense (of course, we could call this feature 'phenomenality' if we want, but I take it that phenomenal realists will not want to do that). Now, the challenge of the target article was to articulate a concept of phenomenality that is recognizably substantive (and so not compatible with illusionism) yet stripped of all commitments incompatible with physicalism. Schwitzgebel hasn't done this, since his conception is not substantive.

Nevertheless, Schwitzgebel has succeeded in something perhaps more important. He has defined a neutral explanandum for theories of consciousness, which both realists and illusionists can adopt. (I have referred to this as consciousness in an inclusive sense. We might call it simply *consciousness*, or, if we need to distinguish it from other forms, *putative phenomenal consciousness*.) In doing this, Schwitzgebel has performed a valuable service.

4. Opponents

This section responds to four papers by opponents of illusionism. Each makes important points, which deserve more discussion than there is space for here, but I shall indicate the general lines of reply that I favour. As I stressed in the target article, my aim is not to refute alternative positions but simply to establish the attractions of illusionism. I begin with two commentators who write from a conservative realist perspective.

Katalin Balog mounts a forthright defence of common-sense phenomenal realism. In particular, she argues that explanatory gap considerations do not give physicalists reason to prefer illusionism, since they can be explained by a version of the phenomenal concept strategy. Specifically, Balog proposes that direct phenomenal concepts are partly constituted by the experiences they refer to and refer to them in virtue of this fact: phenomenal states serve as their own modes of presentation. This, she argues, gives us a direct and substantial access to our phenomenal states, which is very different from the access science gives us and creates the impression of an explanatory gap. I indicated my misgivings about this strategy in the target article (see also Tartaglia's commentary in this issue), but I shall add a few more comments here.

First, it is not clear how constitutive self-reference is supposed to work. A concept may be partially constituted by a state without referring to it, and Balog does not explain what further factors are involved in the case of phenomenal concepts. More importantly, at best the account explains why we think phenomenal states could be *non*-physical; it does nothing to explain how they could be physical. To see this, it is enough to note that we might have a concept of this kind for a non-phenomenal state, such as a propositional attitude. The concept might represent the state in a substantial and direct way, opening a potential gap between facts about it and the neural facts (indeed illusionists might accept that phenomenal concepts represent sensory states in this way). Yet we might feel no resistance to the idea that the state represented is physical — nothing in our grasp of it need *rule out* its physicality or make it difficult to conceive of. Yet that is just what we feel about phenomenal states — we can't understand how they could be physical, and the phenomenal concept strategy sheds no light on the matter. Moreover, Balog makes it clear that she thinks that experiences possess introspectable phenomenal character when they are not being targeted by phenomenal concepts:

> Thinking about [an experience] and simply having the experience will then share something very substantial, very spectacular: namely the phenomenal character of the experience. (Balog, this issue, p. 45)

But if we have a grasp of phenomenal character that is independent of our phenomenal concepts, then we cannot explain away its puzzling features by reference to those concepts. Simple acquaintance will be sufficient to raise all the familiar questions. What is this 'very substantial, very spectacular property'? How do brain processes generate it? Why is it not detectable from other perspectives? And how can we be aware of it when we are not representing it to ourselves?

Balog raises other objections to illusionism. It is, she says, 'utterly implausible' (p. 42). It 'flies in the face of one of the most fundamental ways the world presents itself to us' (p. 47) and manifests a misguided — and perhaps dangerous — scientism (p. 42). I am not unsympathetic to Balog's worries, but I think they are unfounded. As we have seen, illusionists do not deny the existence of consciousness in the innocent sense defined by Schwitzgebel; they merely offer a different account of its nature. And here, as Balog puts it, 'the question comes down to the epistemic authority accorded to introspective awareness vs. scientific theorizing' (p. 47). In developing a theory of consciousness, I plump for the latter. We have abundant evidence of the unreliability of introspection, and there is no reason why an evolved cognitive system should represent its internal states to itself in a transparent way, as opposed to an adaptively useful one. I do not accept that this view manifests a scientistic attitude. My motive for adopting it is the same as that for relying on scientific theorizing to explain any other aspect of the natural world — namely, a desire to have an account of the phenomenon that is as far as possible undistorted by human interests and biases. But seeking such an account need not involve dismissing or devaluing other ways of describing the world, including folk theories, the humanities, the creative arts, and spiritual traditions, all of which may pick out patterns and capture insights that are not tractably expressible in the language of science. Nor does endorsing illusionism require us to give up the language of phenomenality: we may continue to employ it as a useful way of characterizing our inner life, while recognizing that it is, in Quine's phrase, an essentially dramatic idiom (Quine, 1960, p. 219). (In comparing qualia to fictions, I am not expressing a negative view of qualia so much as a positive one of fiction.)[6]

---

[6] Balog suggests that I take a 'negative view' of qualia, as indicated by my use of the term 'embarrassed' in reference to them (Balog, this issue, p. 47). In fact, I used the term to refer to the difficulty *realists* face in accounting for the potency of consciousness — a difficulty illusionists avoid.

Balog also claims that I illicitly appeal to qualitative properties in presenting the case for illusionism. One of her worries concerns introspective phenomenal concepts. If illusionism is true, she points out, these will be either universally misapplied or meaningless. In the former case, it will be miraculous that we have them, and in the latter they will be merely 'mental junk'. It is, she suggests, only because we are already acquainted with phenomenal properties that we can make sense of our having introspective representations that refer to them.

I accept that providing a theory of content for phenomenal concepts will be a major challenge for illusionism, but I don't think we have reason at this stage to write it off as unsurmountable. Balog argues that the problem is particularly hard because phenomenal concepts, unlike other non-referring ones, are simple and direct ones, with no compositional structure. I suspect this is wrong and that the apparent simplicity of phenomenal concepts belies a lot of structure, which we shall tease out as we learn more about the mechanisms involved. (For example, it is plausible that phenomenal concepts contain a distinct affective component. Consider sufferers from pain asymbolia, who recognize pains but no longer find them unpleasant or distressing. Is their introspective concept of pain the same as ours, or a thinner one, which lacks an affective dimension?) As I indicated in the target article, illusionists can appeal to a wide range of factors to explain phenomenal content, including conceptual links, links to non-conceptual sensory and introspective representations, and recognitional capacities for neural states, and they can explain our acquisition of phenomenal concepts as due to developmental processes, individual theorizing, cultural transmission, or a combination of all three.

It may be the case that phenomenal concepts do not pick out metaphysically real (albeit uninstantiated) properties (I suspect they embody inconsistent theoretical commitments, in which case they presumably do not). But even if so, it does not follow that they are simply 'mental junk' as Balog puts it. They may still play an important role in expressing our relation to stimuli and orienting us with respect to our own sensory processes.

I grant that it is not easy to see how this approach can account for the apparent richness of our phenomenal worlds, but it is not clear that realists are much better placed to do this. Would the mere existence of a causal connection to a phenomenal property account for the richness of our conception of it? Balog may say that it is our direct acquaintance with phenomenal properties that gives content to our phenomenal concepts. But this is not a genuinely explanatory move, since the acquaintance relation itself is wholly unexplained (if anything, it is the assumption of acquaintance that is illicit in this context). My sketch of an illusionist theory of phenomenal content may be, as Balog puts it, hand-waving, but it is at least hand-waving in the direction of a coherent research programme.

Balog's other worry concerns the functional sense of 'what it is like', which I introduced in contrast to the phenomenal one. To say that one's experiences are *like something* in this functional sense is to say that one has information about them, provided by functionally defined representational mechanisms. We have a strong intuition that such informational processes would not be sufficient to give us inner lives of the kind we have, but I suggested that if we had a richer and more detailed account of the representations involved, then we might lose this intuition. Balog objects that in suggesting this I am proposing a functional-representational analysis of what-it's-like-ness, which is not only highly implausible but a form of conservative realism rather than illusionism. This mistakes the suggestion, however (which was perhaps not clearly expressed). The idea was not that our *current* notion of what-it's-like-ness is a functional one. I think it is not. Rather, it was that as we develop a richer understanding of the representational processes involved in introspection, we may *reconceptualize* our inner lives in terms of the functional notion of 'what it is like', coming to see them as constituted by representational processes that create the illusion of phenomenality. This wouldn't vindicate the reality of what-it's-like-ness in the phenomenal sense, any more than coming to see a magician's performance as a series of deceptive manipulations would vindicate the reality of the apparent effect.

In his commentary, **Jesse Prinz** argues that illusionism, while not absurd, is less attractive than reductive realism (he dislikes the term 'conservative'), which identifies phenomenal properties with functional or physical ones. He suggests that illusionists, like dualists, expect too much from a physicalist theory of consciousness: identities are not deducible but inferred from correlations and partial explanations, and if a reductive theory can explain enough of the features of consciousness, then we are justified in adopting it.

I take this challenge seriously. Prinz has done tremendously important work in identifying the psychological and neural correlates of consciousness and in providing the sort of partial explanations that he thinks are the best we can hope for in this area (see, in particular,

Prinz, 2012). But while I do not deny that we may be able to provide reductive accounts of many aspects of conscious experience, I doubt that these will be sufficient to justify realism about phenomenal properties in anything like the traditional sense.

My worries centre on explanatory gaps. While identities may be initially inferred on the basis of partial explanations, we expect to be able to render them intelligible, giving reductive explanations of higher-level properties in terms of more basic ones. Why should consciousness be an exception, especially when the feature that resists explanation is such a central one? A partial explanationism that doesn't explain phenomenality itself is *too* partial.

Prinz suggests that gaps arise in the case of consciousness because we have direct acquaintance with phenomenal properties. We know our experiences by having them, not by representing them, and this gives us a special sort of knowledge, phenomenal knowledge, which science cannot provide. In the target article I dismissed the notion that physical states can directly reveal themselves to us, but Prinz argues that we can explain it in processing terms. The neural states that constitute conscious experiences directly reveal themselves to us in virtue of being accessible to higher cognition. They do not need to be represented; they are themselves representations and constitute our awareness of external properties. Yet they also tell us something about themselves and the subjective way they present the world to us.

This is an ingenious move, which deserves detailed consideration, but my first response is sceptical. I see how the accessibility of a representational state would give us a grasp of the worldly properties it represents, but I fail to see how it could give us any knowledge of intrinsic properties of the state itself. (Of course, those who adopt a representational theory of consciousness will say that grasping the content of an experience just is grasping its phenomenal character. But then phenomenal knowledge would have no *distinctive* content. I assume Prinz does not want to take this line: indeed, he thinks that phenomenal properties are intrinsic ones; this issue, p. 191.) Prinz argues that conscious states inform us about themselves because they present the world to us in a subjective way, reflecting categories and divisions imposed by our minds, such as categorical colour boundaries. But, arguably, this is just to say that experience *misrepresents* the world in certain ways, and it is unclear how access to a misrepresentation of the world can afford us any knowledge of the representing state's intrinsic nature. Moreover, even if acquaintance did give us knowledge of intrinsic phenomenal properties of neural states, I do not see how this knowledge could have any cognitive significance for us. Neural states affect cognitive processing in virtue of having causal properties that correlate with their representational content. Other causal properties they may possess can have no distinctively *cognitive* effects. So, if phenomenal features are not themselves represented, then they cannot have cognitive effects, even if they have effects of other kinds. This is all very brief of course, but there is a strong case for thinking that all knowledge of the physical world, including those parts of it that constitute our own minds, is representationally mediated.

Prinz also makes some points against illusionism, which he thinks is prone to collapse into reductive realism. His first point concerns the reference of phenomenal terms. We learn these terms, he argues, by pointing to examples, not by description, so if the exemplified states have physical correlates, then it is to these that the terms refer, and realism is true. My response is that a term may be acquired by pointing yet also have a descriptive component. We may learn phenomenal terms from examples but conceptualize their referents as phenomenal in the sense created by the phenomenality language game (this may involve a later theoretical accretion to concepts originally defined ostensively). If, as illusionists claim, this conception radically misrepresents the states referred to, then it is misleading to use phenomenal terms for them. Of course, if Prinz is right, then the phenomenal conception does *not* misrepresent those states; but this just takes us back to the issue of whether reductive realism is true.

Prinz's second point concerns the nature of illusions. Illusions require seemings — representations of the illusory situation. What should illusionists say about the seemings that constitute phenomenal illusions? If they say they are phenomenal states, then they have not eliminated phenomenality, but if they say they are beliefs, then they cannot explain the apparent richness of experience, since experience is more fine-grained than belief. I have already discussed this issue in various places, so I shall be brief. I suspect that beliefs may do a lot more work here than we imagine, but the illusionist need not claim that they do it all. Phenomenal illusions may depend on representational states that are fine-grained but not phenomenal. They may, for example, be parasitic on sensory representations themselves (functionally defined), arising when these representations are targeted by conceptual mechanisms. (The thought is that we might introspectively represent a phenomenal property via a sensory content, as the phenomenal property with *this kind* of content.) Or they might depend on quasi-perceptual introspective mechanisms, which give us the sort of caricatured access to our own neural processes that I discussed earlier. These fine-grained representations might be bound up with various beliefs, creating multi-faceted introspective states. This again is hand-waving, but I think it is enough to divert *a priori* objections to the illusionist project.

Prinz concludes his comment with some thoughts on the illusion problem. Explaining the illusion of phenomenal consciousness, he suggests, may be no easier than explaining phenomenal consciousness itself and may require very similar resources, in which case the attractions of illusionism diminish. I don't wish to play down the hardness of the illusion problem, but I think this overstates the case. As a rule, the more magical and inexplicable something seems, the easier it is to create the illusion of it than the reality, so the very hardness of the hard problem should give us reason to think that the illusion problem will be easier to solve. If we resist this conclusion, it may be because of the familiar intuition that there is no appearance/reality distinction for consciousness — the illusion would have to be as magical as the reality. I have already argued that we should not trust this intuition, but I want to add a further point.

Coming to see consciousness as an illusion may involve not only theorizing about the mechanisms involved but also reconceptualizing our inner lives. Compare stage magic again. Working out how a magic trick is done involves resisting our natural interpretation of what we see — our intuitive sense of what forces are at work, what causal sequences occur, what properties items have, and so on. In doing this, we can achieve a sort of aspect switch, reconceptualizing the events we see as a sequence of clever manipulations rather than a simple miraculous effect. In order to understand consciousness, we may need to achieve a similar aspect switch in introspection, reconceptualizing our inner lives as constituted by complex multi-dimensional representational processes rather than simple phenomenal effects. The test will be what happens as we learn more about the psychology and neurophysiology of consciousness and try to apply its findings introspectively. I suspect we shall find that the illusion problem seems increasingly tractable. Of course, we may still default to the intuitive phenomenal aspect, and when we do we may still feel the pull of the hard problem; but this psychological fact will seem increasingly irrelevant.

I turn finally to two commentators who adopt a radical, non-physicalist view of consciousness. **Philip Goff** does not attack illusionism directly but challenges one motivation for it — the claim that radical realism is incompatible with our scientific worldview. The target article offered some reasons for thinking that there is a tension here, but Goff argues that these are not compelling. Radical realism, he argues, is not inconsistent with third-person science, especially as realists can adopt a Russellian monist view, which identifies phenomenal properties with absolutely intrinsic properties of physical entities and systems (this view is sometimes characterized as a form of physicalism, but it is a non-standard one, and it is a radical position in my sense). And although radical realism involves *additional* metaphysical commitments, beyond those of third-person science, Goff argues that these are not unacceptable. We have introspective grounds for making them, and their disadvantages are not overwhelming.

In reply, I concede that radical realism need not be strictly inconsistent with science, at least if it takes a Russellian monist form (interactionism is a different matter, though there is doubtless a lot more to be said on the matter). If phenomenal properties are absolutely intrinsic ones, then they are simply invisible to third-person science. Moreover, I do not think it is incoherent to suppose that we might supplement our scientific picture of reality in the way Goff proposes, though it might involve some rather extravagant metaphysical commitments.[7] (This isn't to say that I think Russellian monism is without internal problems. In particular, it faces serious problems in explaining how and when subjects of consciousness combine and how subjects correspond to physical organisms.) The case for illusionism is not that there are no alternatives, but simply that it is much *better* than the alternatives — more economical, more elegant, and, most importantly, more explanatory. This is not the place for an assessment of Russellian monism, but I shall say a few words about the last point — explanatory power.

Goff writes as if Russellian monism offers a new framework for research on consciousness, freed from the constraints of what he calls *radical naturalism*. But it is difficult to see how it could do this. Since the theory treats phenomenal properties as intrinsic ones, it offers no predictions as to the behaviour of physical systems (this is what ensures its consistency with third-person science), and its data, which are introspective episodes, cannot be intersubjectively compared and

---

[7] Goff himself argues for a *cosmopsychist* form of Russellian monism, according to which the universe itself is conscious (Goff, forthcoming).

checked. There is no basis here for a collective science of consciousness, but, at best, for multiple individual sciences. Indeed, even this is too optimistic. The data for a radical realist science are *immediate* introspective episodes, and we have no way of comparing such episodes over time or checking that our beliefs about past episodes are accurate. These are not merely 'methodological difficulties', as Goff puts it (this issue, p. 91), but in-principle obstacles to a radical realist science of consciousness. Russellian monism gives us no new explanatory purchase on the world but merely adds a fifth wheel, which serves no function other than to underwrite our conviction that we have direct introspective access to phenomenal properties. Perhaps if all attempts to explain consciousness within the standard scientific framework were to fail, we might fall back on this view. But it would be premature to adopt it now. Radical realism may not be incompatible with third-person science, but it is a poor substitute for it.

I shall add one further, minor comment. Goff asks why I assume that consciousness must be known with certainty if it is to be a datum; in general, we needn't be certain of our data (Goff, this issue, p. 92). This is true, but in other contexts we are prepared to question our data in the light of theory. If realists wish to insist that the reality of phenomenal consciousness is a *bedrock* datum, which cannot be questioned (which was the suggestion under consideration in the context of my original remark), then I think they do need to claim that it is known with certainty. Of course, if realists accept that the introspective data are open to question, then it is a different matter. But illusionists will be happy to fight on this ground.

Martine Nida-Rümelin begins her commentary by rejecting the widely held view that phenomenal consciousness consists in having experiences with phenomenal properties. Talk of experiences having phenomenal properties is, she argues, a confused way of talking about *subjects* having *experiential* properties, where these are properties that it is like something to undergo. If this is correct, then it immediately undercuts illusionism as originally presented. If it is confused to think that being phenomenally conscious involves having experiences with phenomenal properties, then it is equally confused to think that it involves having experiences that are misrepresented as having phenomenal properties. However, as Nida-Rümelin notes, illusionists may simply recast their view as the claim that we misrepresent ourselves as having experiential properties, and she goes on to argue against this claim. (In fact, she argues that the claim is *necessarily* false. She accepts that phenomenal consciousness does not fit into our standard scientific worldview and that theoretical considerations would favour illusionism, were it a possible view.) Since this is the crux of the matter, I shall focus on her arguments, granting her earlier move for the sake of argument.

Nida-Rümelin's main argument appeals to facts about reference fixing. She argues that reference to experiential properties should be introduced in the following, two-step way. First, we point to paradigm examples, such as suffering pain, feeling sad, or being visually presented with blueness. Second, we establish reference to a feature all the examples share, using metaphors and provisional descriptions (perhaps talking of 'what it is like' to have the properties), but without making any theoretical commitments as to the nature of the feature. This shared feature is what marks out experiential properties and thus phenomenal consciousness. Since illusionists deny the existence of experiential properties, Nida-Rümelin argues, they must either deny that this procedure picks out a common feature of experiential properties or say that experiential properties are never instantiated. Neither option, she argues, is attractive.

This argument is similar to Schwitzgebel's, and my response is similar. I grant that the first step picks out real properties — the personal-level properties we call 'being in pain', 'feeling sad', 'seeing a blue colour',[8] and so on. Illusionists do not deny that *something* is going on when we are in pain or feeling sad. And I grant, too, that the second step establishes reference to a common feature of these properties. However, I deny that it is the sort of feature realists think it is. It is not some intrinsic quality, akin to the property characterized by the phenomenality language game. Rather, it is (roughly) the property of having a cluster of introspective representational states and dispositions that create the illusion that one is acquainted with some intrinsic quality. I am sure that this is not what Nida-Rümelin thinks the procedure picks out, but I don't see how she can rule out the possibility. She makes it clear that in the second step reference is to be fixed by ostension, not description (she says that any descriptions used are merely an aid to identification and may not survive later theorizing — this issue, p. 168). So I am happy to concede the truth of realism about experiential properties *in this sense*. However, this is a very

---

[8] I'm not sure about 'being visually presented with blueness', which is how Nida-Rümelin puts it (this issue, p. 165).

weak kind of realism, which is compatible with the ontology of illusionism.

Nida-Rümelin briefly outlines another argument against illusionism, which turns on the claim that our awareness of experiential properties is direct and unmediated:

> If to have a property P and to be aware of having P is one and the same thing, then the awareness of having P cannot possibly 'misrepresent' oneself as having P. On that view, being aware of having an experiential property by having that experiential property does not involve any further step (no reflection, no introspection, no conceptualization) and therefore leaves no room for any kind of illusion. (Nida-Rümelin, this issue, p. 167)

The appeal here is, of course, to direct acquaintance: experiential properties reveal themselves to us immediately, leaving no room for error. This notion has cropped up frequently in this discussion, and it is a central one. *Pace* Prinz, I don't believe it is possible to give a physicalist account of direct acquaintance in any robust sense. I cannot see how physical properties can reveal themselves in this way to a physical cognitive system (by 'physical properties' I mean structural and dispositional properties of the sort described by third-person science, not absolutely intrinsic ones). So, as a physicalist, I maintain that our sense of being directly acquainted with experiential properties is itself an illusion, an artefact of the way our sensory and introspective systems are structured. This will not convince Nida-Rümelin, of course, whose metaphysical commitments are different from mine. But I think we are close to bedrock here, and that's a good place to stop.

5. Conclusion

I conclude by thanking the commentators once again. They have forced me to think hard about my position and its commitments, but they haven't shaken my belief in it. If it's an illusion to think that phenomenal consciousness is an illusion, then I'm not disillusioned.[9]

---

[9] I am grateful to Daniel Dennett, Eileen Frankish, Nicholas Humphrey, Maria Kasmirli, and Miloš Tomin for comments, advice, and assistance.

References

Dennett, D.C. (1991) *Consciousness Explained*, New York: Little, Brown.

Dennett, D.C. (2005) *Sweet Dreams: Philosophical Obstacles to a Science of Consciousness*, Cambridge, MA: MIT Press.

Dennett, D.C. (2013) Expecting ourselves to expect: The Bayesian brain as a projector, *Behavioral and Brain Sciences*, **36** (3), pp. 209–210.

Dick, P.K. (1995) *The Shifting Realities of Philip K. Dick: Selected Literary and Philosophical Writings*, New York: Vintage Books.

Doyle, A.C. (1892) *Adventures of Sherlock Holmes*, New York: Harper and Brothers.

Frankish, K. (2012) Quining diet qualia, *Consciousness and Cognition*, **21** (2), pp. 667–676.

Garfield, J.L. (2015) *Engaging Buddhism: Why it Matters to Philosophy*, Oxford: Oxford University Press.

Goff, P. (forthcoming) *Consciousness and Fundamental Reality*, New York: Oxford University Press.

Humphrey, N. (2011) *Soul Dust: The Magic of Consciousness*, Princeton, NJ: Princeton University Press.

Pereboom, D. (2011) *Consciousness and the Prospects of Physicalism*, New York: Oxford University Press.

Prinz, J.J. (2012) *The Conscious Brain*, New York: Oxford University Press.

Quine, W.V.O. (1960) *Word and Object*, Cambridge, MA: MIT Press.